PUBLIC HEALTH IN THE 21ST CENTURY

BIOLOGICAL IMPLICATIONS OF HUMAN MOBILITY

PUBLIC HEALTH IN THE 21ST CENTURY

Additional books in this series can be found on Nova's website under the Series tab.

Additional eBooks in this series can be found on Nova's website under the eBook tab.

PUBLIC HEALTH IN THE 21ST CENTURY

BIOLOGICAL IMPLICATIONS OF HUMAN MOBILITY

SLAWOMIR KOZIEL,
RAJA CHAKRABORTY
AND
KAUSHIK BOSE
EDITORS

New York

Library of Congress Cataloging-in-Publication Data

ISBN: 978-1-63485-644-7

Published by Nova Science Publishers, Inc. † New York

CONTENTS

PREFACE

It is widely recognized that social mobility are among the key factors having significant impact on biological wellbeing and health of a population. Social mobility may be of two kinds. Vertical mobility occurs in stratified societies and involves a shift of a person's position (up or down) on a scale of attained education, wealth, power **or** prestige. Vertical mobility may occur between generations, when the shift is measured relative to the position of the subject's parents, or within a single generation, when the subject moves up or down the scale during own lifetime. In recent years, there has been a growing interest in the phenomenon of inter-generation social mobility and its biological and health consequences. Some studies have shown that social factors in early life could influence health status in adulthood. Therefore, socioeconomic status at the time of study is not necessarily the most relevant variable to be taken into account. There are some suggestions that an improvement in social position within the structure of society by an individual relative to the parents may be advantageous for health status. On the other hand, upward social mobility could be selective in case of variables such as body height. It was hypothesized that taller individuals are more likely to move up the scale of educational attainment compared with shorter individuals of the same social background. Another example of vertical mobility is change of social position as a result of marriage to a spouse with a discrepant level of education. In general, when the husband is better educated than his wife, she is described as upwardly mobile; when the wife is better educated than her husband, she is described as downwardly mobile. Significant differences in body built, physical fitness and certain health parameters were found between persons moving up and those moving down the social scale through marriage. Taller women tended to marry men of higher social class, while downwardly mobile men and women were fatter than individuals who ‘marry up’. Women with lesser educated husbands also had a more abdominal fat distribution than comparable women with better educated spouses. Horizontal mobility, in contrast, involves movement either to a different place of residence (e.g., rural-urban, different geographical region) or to a new marital category. Horizontal mobility does not have to cause a shift in the person’s position on any vertical social scale. Human migration is the movement from place to place of individuals or groups of individuals for the purpose of taking up permanent or temporary residence, usually across a political boundary. People migrate for a variety of reasons such as “push” or a “pull” factors. Pull factors are reasons for immigration (moving into a place), such as better educational and occupational conditions, freedom, improved living condition, better food supply, access to medical services etc. Push factors are reasons for emigration (leaving a place), such as food insecurity, war, discrimination, political or religion persecution, violence and armed conflicts, environmental problems etc. The extent of migration in today’s world is enormous.

In most of the developed and much more in developing countries, some features of body size and timing of maturation show clear differences between certain social groups within **a** given society. In most of the auxological studies social groups are defined based on such socio-economical features like level of parental education, occupation, level of income, and demographic features like level of urbanisation at the place of living and family size. The majority of auxological studies published so far definitely confirmed differential influence of social factors on parameters of growth and tempo of maturation. There is also evident social variation in health parameters in adult population.

This well documented differential influence of social factors on biological parameters, such as growth and development of children, health parameters of adults, however, is not always simple to perceive. In view of this, the social factors almost always exert their influences on biological features indirectly. They operate through diverse elements of life style, and living conditions and their changes. Generally it is acknowledged that those elements which depend to a large extent on socio-economic status also include unhealthy habits like smoking and excessive consumption of alcohol, dietary habits, level of hygiene etc. These factors in turn determine the level of morbidity, physical work capacity and level of psychosocial stress. In reality, the ways of action of social factors, might be much more complicated having a network of interrelationships and leaving an additive, synergic or antagonistic effect.

It is worth distinguishing social inequality *per se*, from their biological consequences, which is a subject of anthropological research. It seems that existence of groups with different standards of life in modern societies is irremovable and natural. It could be even postulated that such kind of social inequality provides high motivation for individuals and families toward upper mobility on the scale of social position. It could also be considered as one of the main driving forces of civilisation and emergence of a social group as a whole. However, it is of great importance that different position of social scale does not have a biological consequence in growth and development of children. Outwardly, all developing societies should tent to such a high level of average living standards, with a suitable level of welfare and social care, that differences in social and financial situation do not have biological effects.

The present volume is an endeavour to outlay the possible influence of some important aspects of human migration and social mobility on biological characters of human populations including their health and wellbeing. The book has nine contributions from different researchers working in this area of research. The first chapter written by Budnik and Henneberg demonstrated the effect of social class and mobility on morphological characters of body size like height and body mass index (BMI) in a historical population of Poland. In Chapter II, Chakraborty et al. showed that the time of migration of disadvantaged people to an adverse environment in early period of growth and development may increase health risk in adulthood compared to those migrated after completion of major physical growth period or even compared to those who are born in that environment. Chapter III by J. R. Ghosh revealed the influence of educational and occupational positions on a clinical health parameter, viz., hypertension, among they adult males from the eastern part of India. In the fourth chapter of this volume, S. Ghosh et al. attempted to find out the relationship between the socio-economic status of family and growth of height and weight of school children aged 5-12 in Kolkata, India. Godina et al., in Chapter V, delineated the differences in various anthropometrical measurements in children and adolescents aged 7 to 17 years across different types of schools representing different social strata in Moscow. Chapter VI by Kaczmarek discusses the implications of rural to urban migration and its impact on women's health status in Poland. In the next Chapter, Koziel and Gomula highlight from a study in Poland, the effect of the influence of social mobility of fathers on maturity, measured by the age at menarche, in their daughters. Chapter VIII, by Krzyżanowska and C G Nicholas Mascie-Taylor discusses the impact of regional migration and social mobility on variation in adult height, weight and Body Mass Index; evidence from a British cohort study. In the next chapter, Missoni and Šarac reviewed dietary and lifestyle characteristics in the Eastern Adriatic Islands in Croatia in the backdrop of recent economic transition, urbanisation and migration. The tenth chapter contributed by Singh and Kirchengast compared demographic charateristics, health related and reproductive behaviours between Punjabi women residing in Punjab, and in Vienna, Austria. We hope that this book will be useful for researchers dealing with biological implications of human mobility. It may be of particular interest to human biologists, biological anthropologists, epidemiologists, demographers, economists and other researchers dealing with biological implications of human mobility. We would like to express our gratitude to all the authors for their very valuable contributions. Last but not least, we would like to convey our thanks to NOVA Science Publishers, New York, for all their help and kind cooperation.

Prof. Sławomir Kozieł,
Dr. Raja Chakraborty
and Prof. Kaushik Bose.

In: Biological Implications of Human Mobility ISBN: 978-1-63485-644-7
Editors: Slawomir Koziel et al.

Chapter 1

The Appearance of a New Social Class of Wealthy Commoners in the 19th and the Early 20th Century Poland and Its Biological Consequences

Alicja Budnik[1,*] and Maciej Henneberg[2]
[1]Institute of Anthropology, Adam Mickiewicz University, Poznań
[2]Biological Anthropology and Comparative Anatomy Unit, The University of Adelaide, Adelaide, Australia

Abstract

Introduction. Political and economic changes that occurred on the territory of Poland during the second half of the 19th century contributed to changes in social structures of rural and urban populations. Urban populations underwent a profound change due to industrialisation that caused appearance of new social classes, especially the class of wealthy commoners – the middle class. Influence of social stratification on human morphological characters and on growth at adolescence is well documented for recent populations, however, it is much less studied in the past populations due to scarcity of appropriate data. Objectives. To characterise: 1. body height and body mass of adults and 2. the adolescent growth spurt in various strata/classes of Polish people during the second

* Corresponding Author address: Institute of Anthropology, Adam Mickiewicz University, Umultowska 89. 61-614 Poznań, Poland. Email: ambpp@amu.edu.pl.

half of the 19th century and at the beginning of the 20th century. Material and Methods. Data on body height and body mass of adults (20+ years) and youths (7-20 years) of privileged and poor social strata living in the Kingdom of Poland and in its capital, Warsaw were derived from published historical sources. Means, and where possible standard deviations, of body height, weight and Body Mass Index (BMI) were calculated for various groups. Prevalence of obesity (BMI > = 30kg/m^2) and overweight (BMI > = 25 kg/m^2) was assessed separately for nobility, middle class, working class and peasantry. LOESS curves were fitted to annual cross-sectional increments in body height and body weight of youths of different social classes in order to evaluate the presence and magnitude of growth spurts. Results. The largest average values of body height and body mass of males and females were found for nobility, somewhat lower were values for the middle class. In both groups average BMI exceeded 25 kg/m^2. Peasant males and females were shorter and lighter than those of upper classes. Males and females of the urban working class were even shorter and lighter than peasants. BMI averages of peasants and working class people were below 25 kg/m^2. About 30-40% of nobility and rich middle class people were overweight and 12-21% obese. Among lower classes 5-20% were overweight. Obesity occurred in 4-9% of peasants, but no working class males were obese at all. Children of privileged classes were taller and heavier than children of the working class. Boys of noble and middle class families experienced adolescent growth spurt at 14-15 years with peak height velocity of 7cm/year and weight velocity of 5 kg/year. Among privileged girls adolescent growth spurt occurred at 11-12 years with velocities of 6 cm/year and 5 kg/year. Among poor working class boys and girls there was no adolescent growth spurt. Conclusions. Upward social mobility providing better living conditions and nutrition resulted in increases in body height and body mass of children and adults contributing to the occurrence of overweight and obesity. Adolescent growth spurt was not a universal phenomenon. It did not occur among children of poor social strata. Improved living conditions of privileged classes produced accelerated growth at adolescence.

Keywords: Kingdom of Poland, Warsaw, social strata, adults, children and youths, body height and weight, BMI, prevalence of overweight and obesity, adolescent growth spurt

INTRODUCTION

Several important political, economic and legal changes jointly produced a profound change of social structures of people inhabiting Polish lands in the second half of the 19th century. Abolition of the serfdom awarded freehold of the land to peasants. This gradually changed rural economy towards the free market capitalist model replacing outdated feudal practices. This process progressed at different rates in various parts of Poland occupied by Prussia, the Austro-Hungarian Empire and Russia. In the Kingdom of Poland created by the Congress of Vienna in 1815, that remained under Russian rule, the serfdom was abolished by the tsar's decree only in 1864, however, some changes in the rural economy commenced earlier (Chwalba 2000, Stankiewicz 1973). The reform of agricultural economic relations – positive in the broader perspective - exacerbated economic stratification of the rural population forcing landless workers and small farm owners to seek employment in industrialising urban centres. This phenomenon, although often percieved by individuals involved as a dramatic change, had positive economic effects by freeing rural areas of the unemployed population surplus and supplying substantial labour force to industrialised cities (Chwalba 2000).

Urbanisation and industrialisation initiated formation of new social strata in the cities. One of such new classes were labourers, the proletariat. In large industrial centres of the Kingdom of Poland members of the working class were derived mainly from rural farmhands and peasantry, but some tradespeople and even poorer members of the gentry became wage labourers, too, for example in Warsaw, the capital of the Kingdom (Chwalba 2000). In the public opinion the image of the working class – the proletariat – is that of the poorest, socially and economically disadvantaged social stratum. In reality, economic and social status of wage labourers was dependent on their skills and some of them, especially those highly skilled ones, tended towards middle class lifestyle. Of course, there were many members of the working class who earned poor wages, were unskilled, unemployed and led disorganised lives (Kałwa 2006). New wealthy middle class (burgeousie) comprised financiers, bank owners, proprietors of manufactures, factories and large trading enterprises. These people orignated from various social strata, most often from well-established city dweller families, tradespeople, merchants, but sometimes even from among gentry and landowners. They aspired not only to achieve wealth, but also to improve their social standing and prestige trying to form social relations with landed gentry and aristocracy. These latter classes, though traditionally aloof and hermetic, could not withstand economic competition from the rising middle class and through social contacts with the newly affluent burgeouisie changed their behaviour and even, sometimes, intermarried with newly rich families. There were similar proceses of breaking class barriers in rural areas where poorer members of the gentry mixed with peasants (Chwalba 2000, Kałwa 2006).

Summing up, in the new early industrial urban environment most individuals became members of the working class, but some rose to positions of affluence through their skills at enterpreneurship. Their income became comparable with that of the members of the traditional noble class. The nobility itself was economically differentiated from landless descendants through owners of modest agricultural estates to aristocracy. In rural areas differences between peasantry and poor nobility became blurred.

Since living conditions are dependent on income and also on the education level and social milieu, mobility from one social stratum to another influenced growth and development of children and biological characteristic of adults.

Influence of socioeconomic differences on such traits as body height and weight, weight/height ratios and menarcheal age is very well documented for various populations living during the last 100 years (Eveleth and Tanner1990, Bogin 1988, 1999; Ulijaszek et al. 1998, Malina and Bouchard 2004). In Poland, extensive studies were conducted by the Department of Anthropology of the Polish Academy of Sciences in Wroclaw and by a number of other human biologists. They studied relationships between variously measured social stratification and body height, weight, BMI, adolescent growth spurt and menarcheal age in various Polish populations (eg. Bielicki 1981, Bielicki and Welon 1982, Bielicki et al. 1981, 1986, 1988, 2003; Bielicki and Waliszko 1992, Bocheńska 1972, Kaczmarek 1995, 2001, Kozieł et al. 2006, Welon et al. 1983). These authors also studied time trends in described characteristics. Such trends are generally attributed to changing socio-economic situations. Changes of body height, body mass, and the relationships with living conditions in various populations during the 19th century were studied, among others, by Arcaleni (2006), Baten and Blum (2012), Bodenhorn (2010a, 2010b), Cardoso and Caninas (2010), Carson (2005, 2007, 2009, 2011, 2013), Coclanis and Komlos (1995), Cuff (1993), Helmchen and Henderson (2004), Henderson (2005), Hiermeyer (2010), Komlos (1985, 1987, 1990, 2006, 2007), Maloney and Carson (2008), Nicholas and Steckel (1991), Nowak and Piontek (2008), Staub et al. (2010), Sunder (2004), and in Poland by Czapla and Liczbińska (2014), Kopczyński (2006, 2007), Nowak (2011) and Welon et al. (1983).

In relation to historical (pre-World War One) populations similar analyses are limited and difficult to carry out due to shortage of records and to unorthodox ways in which early observations were published. Relationship between social stratification and body size in Poland in historical times has been studied mostly in data collected from conscripts that could be subdivided into rural and urban, or by father's occupation. Results for the Polish territories occupied by Prussia during the 19th and early 20th centuries were given by Nowak (2011) while Kopczyński, using data from the Russian part of Poland for 1866-1913 has shown positive influence of higher educaton on the physical status of conscripts. Those who were university students, skilled technicians and clerks were significantly taller than uneducated, unskilled people. Among less privileged people simple ability to read and write was related to better physical development. This ability improved understanding of the world and also allowed to have somewhat better jobs. This relationship between affluence and body height of conscripts has been also confirmed by a reasonably high level of correlation between stature and an index of income from agricultural production by peasant farms in the Kingdom of Poland (Kopczyński 2006).

Reconstruction of relationships between social stratification and body build and the influence of social mobility on these relations is difficult due to the incompletness and specific quality of historical materials pertaining to the territory of Poland. There are practically no studies on the child development in the past and on its socio-economic determinants. Therefore the aims of this work are as follows:

The Aims

1. Reconstruction of the body size and degree of fatness of adults in relation to their social stratum, affluence and living conditions in the territory of Poland during the later part of the 19th century and at the beginning of the 20th century.
2. Assessment of body height and weight of children and youth of various social strata and evaluation of the presence of the adolescent growth spurt among them in populations of the Kingdom of Poland during the later part of the 19th century and at the beginning of the 20th century.

Methods

Materials used here are two kinds of data derived from published historical sources (Bochenek 1906, Kosieradzki 1905, Miklaszewski 1912, 1914, Rutkowski 1900, 1906). The first kind are data on body height and body mass of adults (>20 years) from privileged and poor groups of Polish people living in the Kingdom of Poland – a part of Poland under Russian rule until the World War One. Most of these data came from the capital city – Warsaw. Altogether information on 3774 individuals was analysed. Among those there were 467 members of nobility (313 males and 84 females), 830 members of the middle class (710 males and 120 females), 990 proletarians (298 males and 692 females) and 1283 peasants (1083 males and 200 females).

The second kind of data comprises published in 1887-1914 (Miklaszewski 1912, 1914) results of cross-sectional studies of height and weight of children and youths 7-20 years old from Warsaw. These data pertain separately to privileged classes (nobility and wealthy middle class (N = 2126, including 706 boys and 1420 girls) and to poor proletariat (N = 990, including 298 boys and 692 girls). Altogether 6260 persons were studied.

For all groups we have considered arithmetic means, and where available, standard deviations. Body mass was often measured in Russian tsar pounds (1 kg = 2.442 pounds). Due to the peculiar presentation of the material, standard deviations were often calculated from grouped data distributions (Jasicki et al. 1962). From body heights and weights we have calculated values of the BMI. Using those, individuals were categorised into normal weight (BMI < 25.0 kg/m^2), oveweight (BMI $\geq$ 25 kg/m^2) and obese (BMI $\geq$ 30 kg/m^2). Where individual data were not available, distributions of BMI values were estimated from normal distributions by taking surfaces under appropriate fractions of the distribution and their corresponding probability densities.

For children and youth, changes of body height and body mass with age were observed. Annual increments of height and weight were calculated from cross-sectional averages. Distributions of annual increments were smoothed using LOESS function of SPSS 22.0 similar to the approach used by Walker et al. (2006).

Results

Figures 1 and 2 present average height and weight of adult males and females in various social strata of the Kingdom of Poland during the late 19th and early 20th century.

As can be seen from the Figures, both biological characters were the largest among nobility. Especially striking is the tallness of noble males. It has been the greatest among aristocrats, the tallest of whom stood 1.91 m high, while the arithmetic mean for the entire group (n = 13) slightly exceeded 1.77 m (Miklaszewski 1914). Differences between nobility and middle class are smaller in body mass than in body height. This is a result of middle class people catching up in terms of living condittions with the nobility during their adult lives. Some of the middle class people bought insurances for 30 thousand roubles (roughly equivalent to $500,000 in present-day currency) while 5 thousand roubles insurance was already seen as indicating significant affluence (Miklaszewski 1914). Upward social mobility of successful middle class people manifested itself in overconsumption of food. Hence, while among the nobles there were practically no intergenerational differences in body mass, middle class people of new (younger) generation were 5 kg heavier than in the older generation (own calculations based on Miklaszewski 1914). This overconsumption by successful middle class people was also reflected in BMI. Average BMI values of the bourgeoisie were the same as those of nobility (Figure 3).

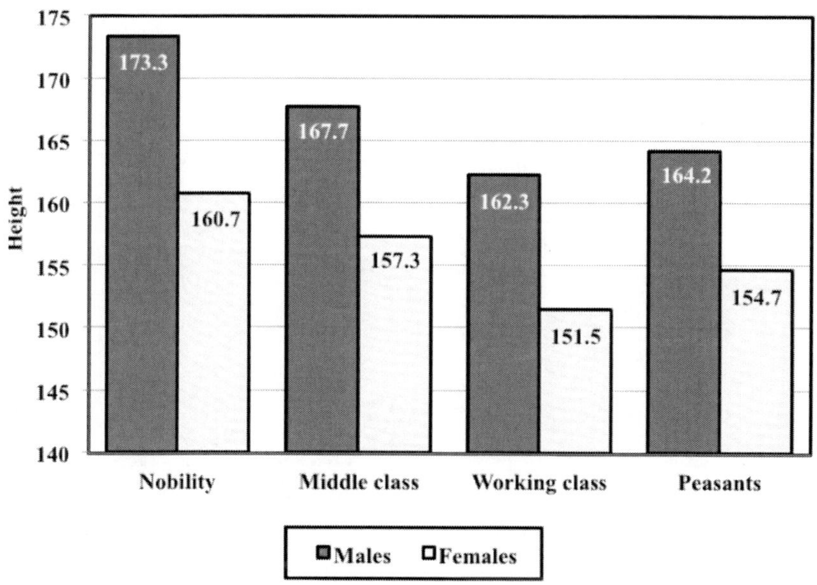

Figure 1. Average body height of adult males and females in various social classes in the Kingdom of Poland in the late 19th and early 20th century.

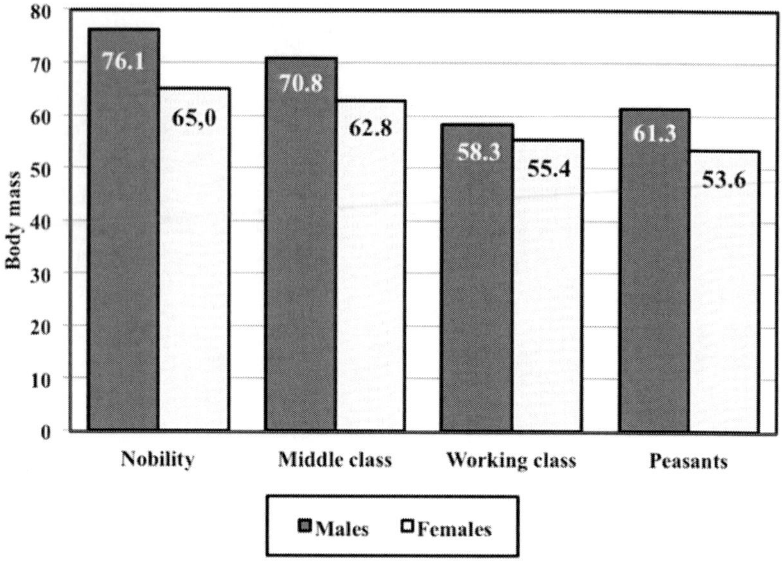

Figure 2. Average body mass of adult males and females in various social classes in the Kingdom of Poland in the late 19th and early 20th century.

Figure 3. Average values of the Body Mass Index in various social social classes in the Kingdom of Poland in the late 19th and early 20th century.

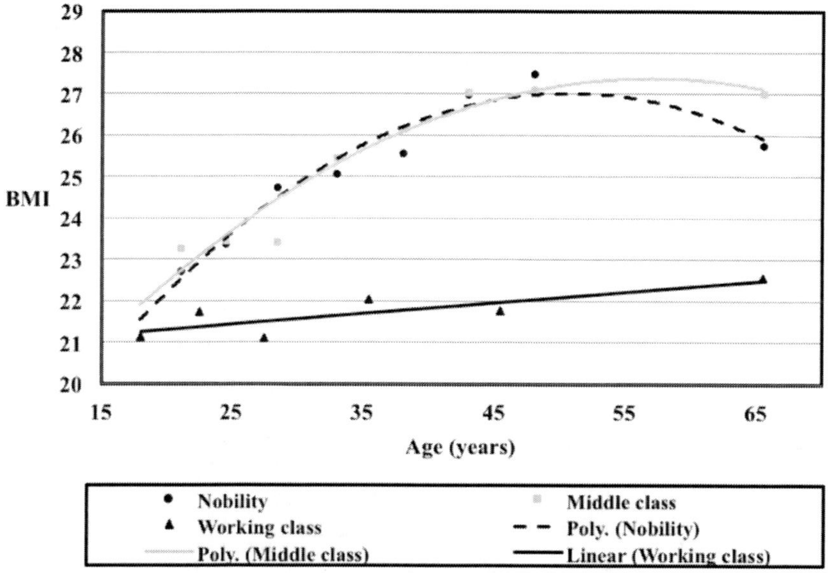

Figure 4. Changes of Body Mass Index with age of males in various social classes in the Kingdom of Poland in the late 19th and early 20th century.

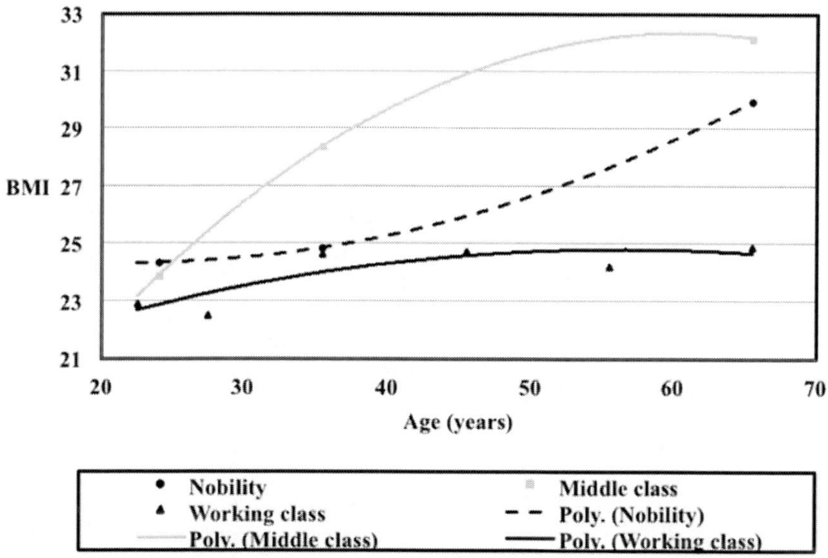

Figure 5. Changes of Body Mass Index with age of females in various social classes in the Kingdom of Poland in the late 19th and early 20th century.

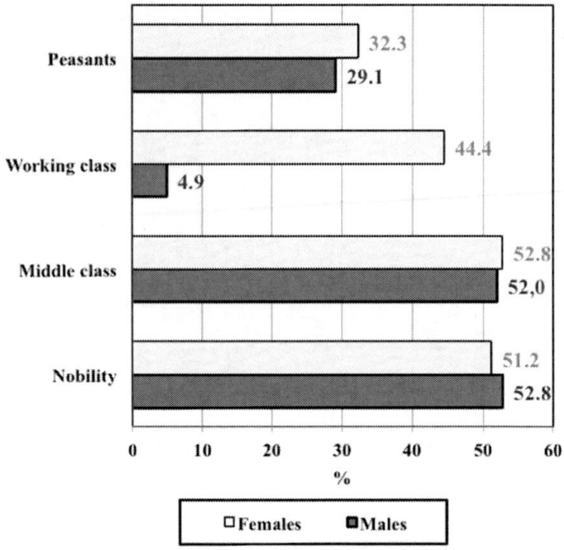

Figure 6. Prevalence of overweight in various social classes in the Kingdom of Poland in the late 19th and early 20th century.

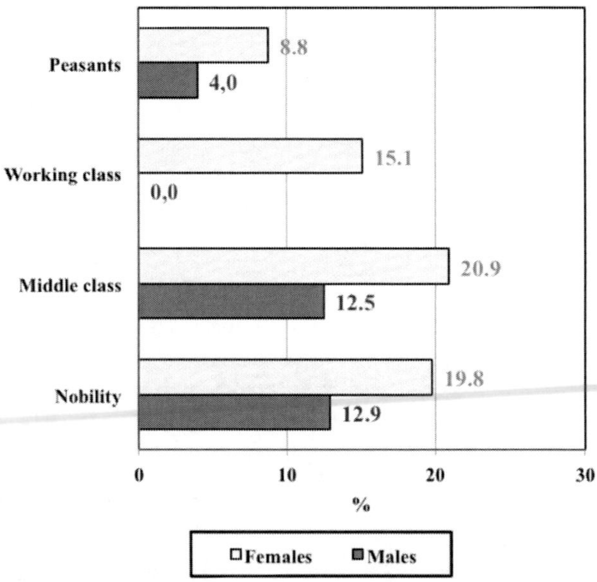

Figure 7. Prevalence of obesity in various social classes in the Kingdom of Poland in the late 19th and early 20th century.

The shortest body height, and in males also the lightest body mass and BMI, were observed among the working class. Proletarian males exhausted by physical labour in very unhygenic conditions maintained low values of BMI practically throughout all their lives – there was no increase of BMI with age (Figure 4). This remains in contrast to working class females who usually stayed at home, and being much less mobile, increased their BMI values towards overweight and obesity after 30th year of life. Their maximal weight exceeded 80 kg while in over 50 years old it was even 90 kg (Miklaszewski 1912). Among proletarian females aged over 30 years in Warsaw average BMI values emphatically exceeded 24 kg/m^2 (Figure 5). However, working class males and females had clearly lower average BMI values in all age categories than nobility and middle class. It is striking that among middle class women BMI values increased with age more than among noble women (Figure 5). This is mostly due to the presence among > 30 years old (29%) middle class females of very high BMI values.

All this resulted in 30% to 40% of the economically privileged classes adults being overweight and 12% to as much as 21% being obese (Figures 6-7). None of the working class males were obese, while not quite 5% of them were overweight. It is interesting that nutritional status of peasants from rural areas surrounding Warsaw was quite good – about 20% of them were overweight and 4% to nearly 9% were obese (Figure 6-7).

The greater body mass of rural people in comparison with urban ones has been already noted by Nowak (2011). Better nutrition of rural people could have been a result of greater efficiency of agricultural methods supported by developing technology. Better agricultural technology abolished putting some lands to lay fallow – in the western part of the Kingdom of Poland at the end of the 19th century fallow lands constituted only 10% of the total arable land. Agricultural production increased. For example, demand for potatoes has been completely satisfied by the end of the 19th century. Some influence could be also ascribed to the increase of the number of peasant-owned farms after 1864 (land reform). Landless rural workers disappeared for a spell (Chwalba 2000).

Table 1 illustrates differences in the consumption of selected foods among urban and rural populations of the Kingdom of Poland.

It can be seen that urban middle class consumed clearly more milk, meat, butter and eggs than peasants. Peasants compensated those deficiencies, to a certain extent, by enormous consumption of potatoes, cereals and legumes. Differences in living standards are also evident in accommodation of families. Table 2 shows percentages of nobility and labourers living in different quality quarters in Warsaw at the end of the 19th century. Nearly ¾ of proletarian families lived in single-room flats, poorly equipped, crowded and situated below the ground level. Only a small proportion of skilled labourers could afford better accommodation. Nobles lived in large, often 7-room, apartments located conveniently on the first or second floor of apartment buildings (Miklaszewski 1914).

Uneven access to consumer goods resulted in pronounced differences in the development of children from different social classes. We have practically no information about the growth of Polish children over 100 years ago. The only useful data come from unique publications of Dr Walenty Miklaszewski (1912, 1914). Even those works barely fit present-day criteria of sound statistical representation. They required special re-analysis. Here we present our own calculations of weighted averages of body height and weight of boys and girls in consecutive age categories. Children were measured in various schools catering to economically privileged strata of Warsaw's population – mainly nobility and affluent middle class. Here we have considered these two classes jointly. Children of poor labourers were measured in summer camps organised by charities and in several schools mostly frequented by working class children. Figures 8-11 present age changes of body height and weight of privileged and poor boys and girls.

Table 1. Consumption of selected food products per person per year in the Kingdom of Poland at the end of the 19th century. Data in kg, own calculations based on Prus (1886)

Food Products	Middle Class	Peasants
Cereals	156	135
Legumes	-	18
Milk	112	56
Butter	8	-
Pigfat	5	3
Meat	37	5
Eggs	130	-
Potatoes	240	426

Table 2. Percentages of persons living in apartments of different size and location in apartment blocks in Warsaw at the end of the 19th century. Own calculations based on Miklaszewski (1914)

Class	Number of rooms				Location in the building	
	1	2	3	4-7	Ground/underground	Higher floors
Working class	74	24	1.6	0.2	83	17
Nobility	-	4	25	71	12	88

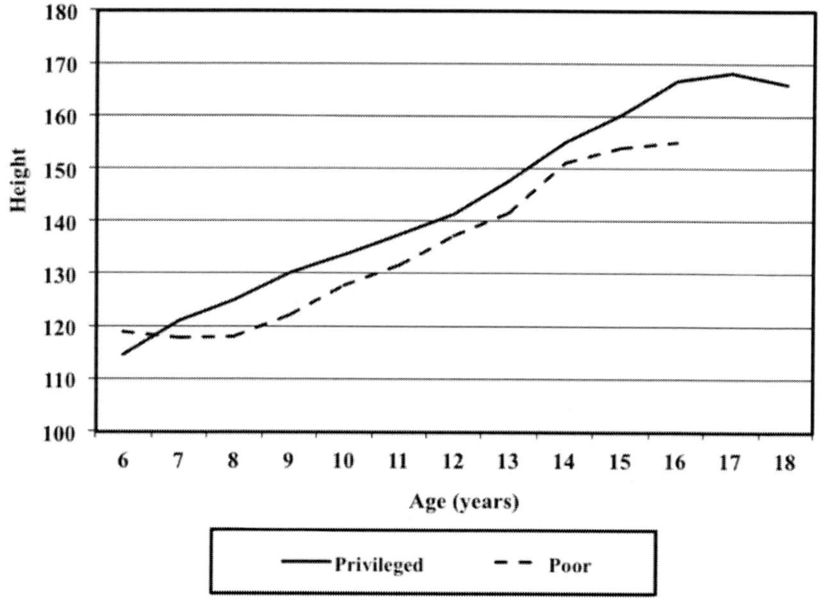

Figure 8. Changes in body height of boys in various social classes in the Kingdom of Poland in the late 19th and early 20th century.

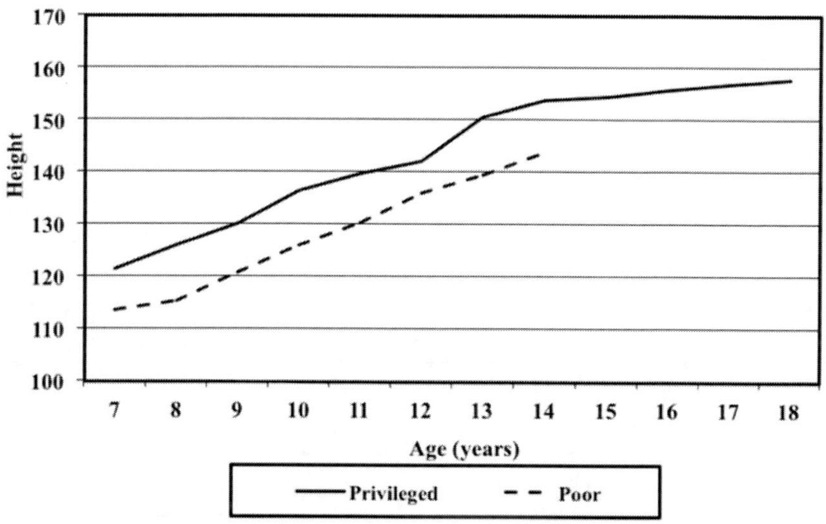

Figure 9. Changes in body height of girls in various social classes in the Kingdom of Poland in the late 19th and early 20th century.

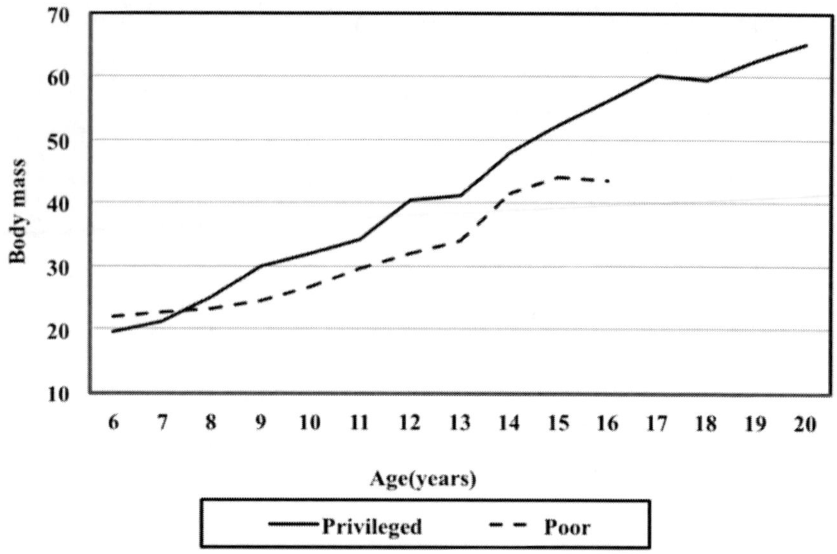

Figure 10. Changes in body mass of boys in various social classes in the Kingdom of Poland in the late 19th and early 20th century.

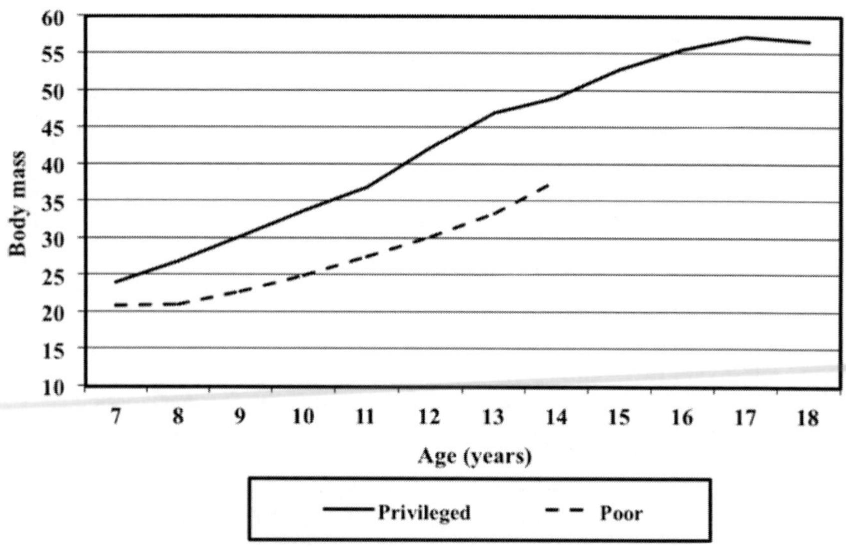

Figure 11. Changes in body mass of girls in various social classes in the Kingdom of Poland in the late 19th and early 20th century.

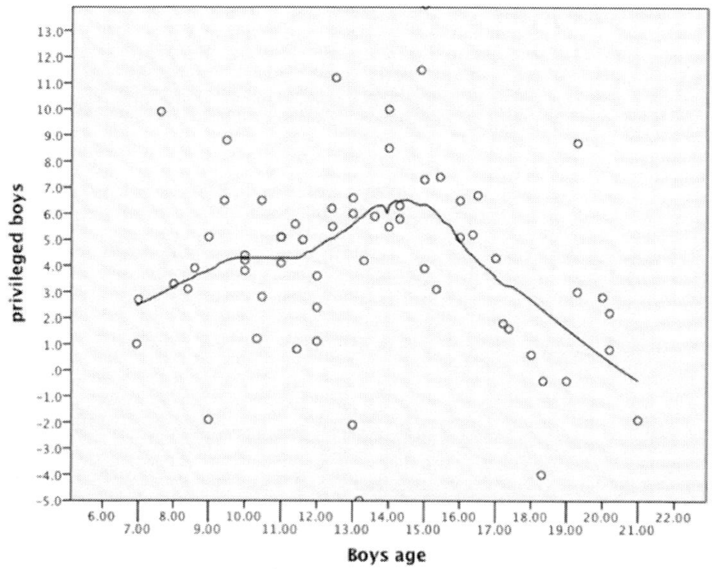

Figure 12. Annual increments of body height of privileged boys in the Kingdom of Poland in the late 19th and early 20th century.

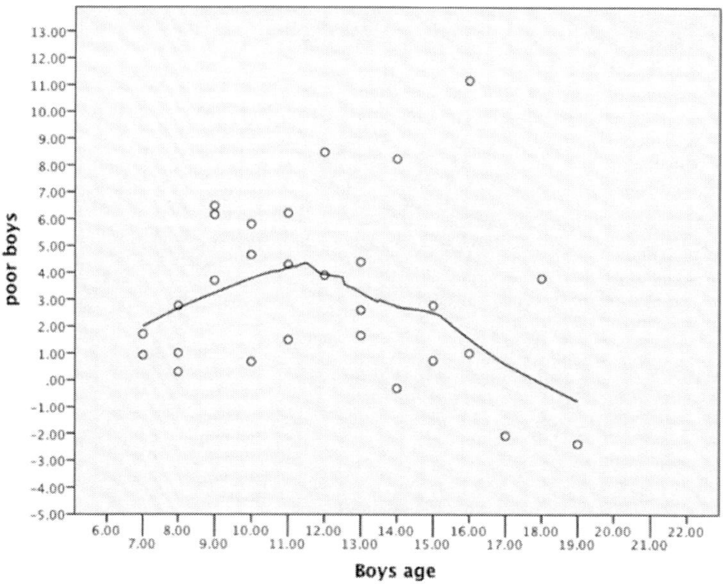

Figure 13. Annual increments of body height of poor boys in the Kingdom of Poland in the late 19th and early 20th century.

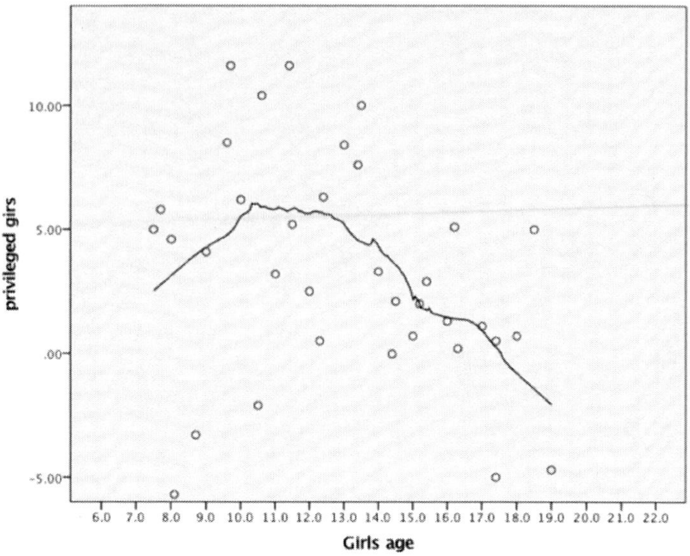

Figure 14. Annual increments of body height of privileged girls in the Kingdom of Poland in the late 19th and early 20th century.

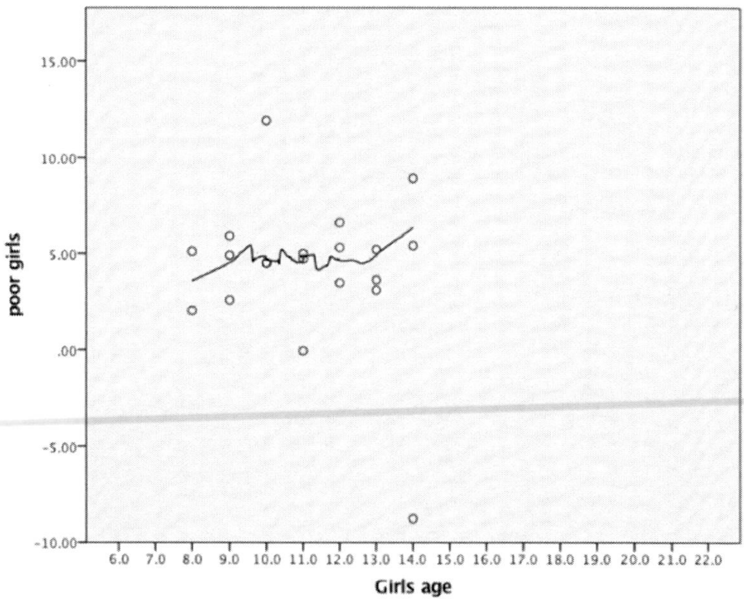

Figure 15. Annual increments of body height of poor girls in the Kingdom of Poland in the late 19th and early 20th century.

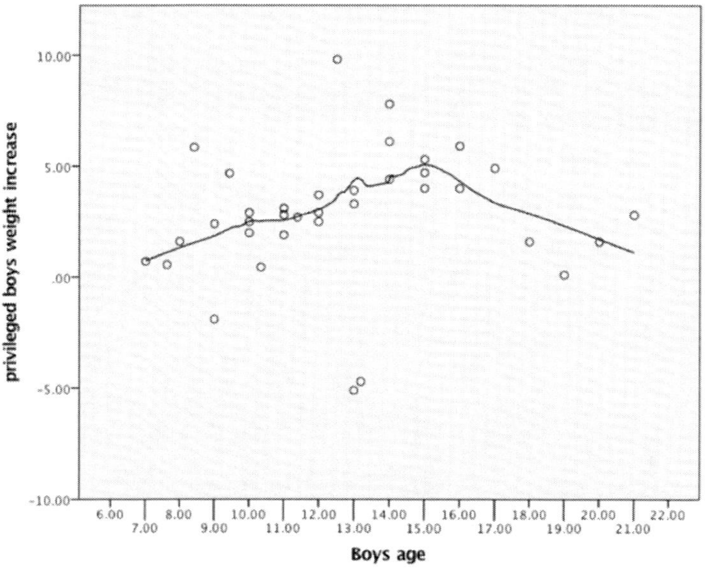

Figure 16. Annual increments of body mass of privileged boys in the Kingdom of Poland in the late 19th and early 20th century.

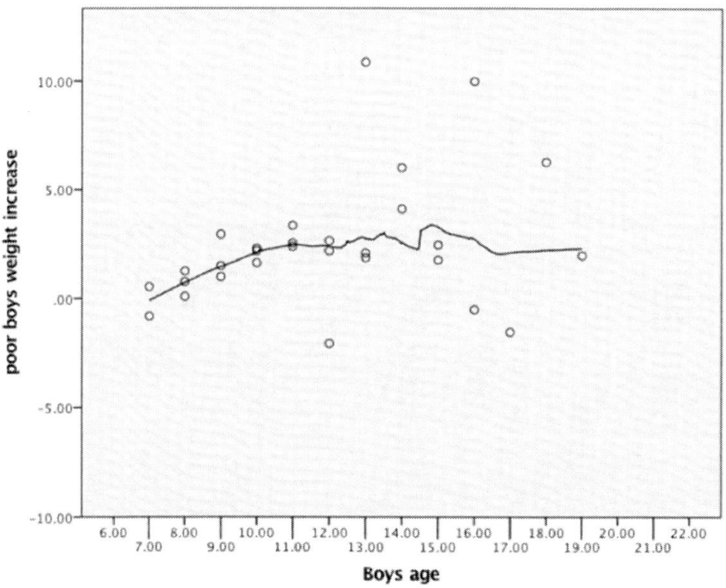

Figure 17. Annual increments of body mass of poor boys in the Kingdom of Poland in the late 19th and early 20th century.

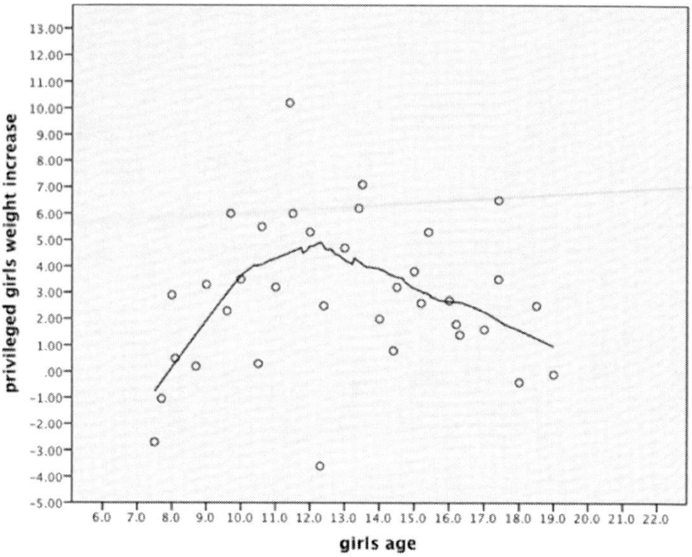

Figure 18. Annual increments of body mass of privileged girls in the Kingdom of Poland in the late 19th and early 20th century.

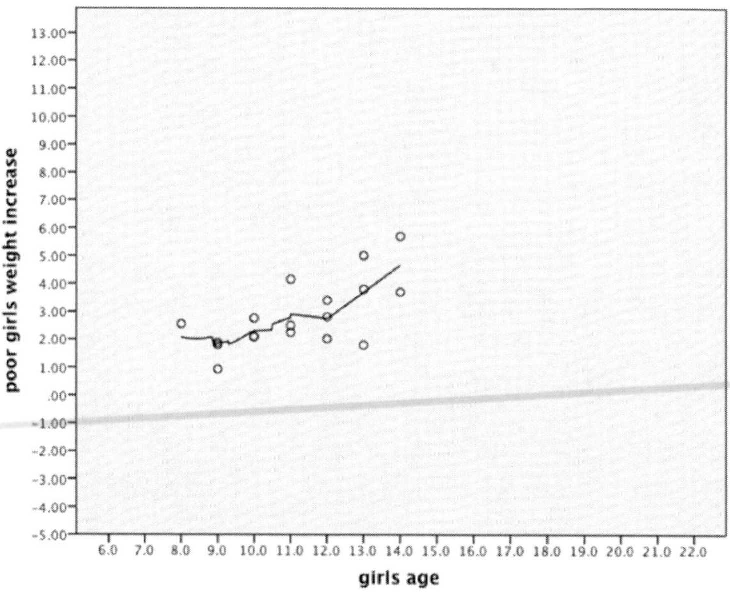

Figure 19. Annual increments of body mass of poor girls in the Kingdom of Poland in the late 19th and early 20th century.

It can be seen that privileged children in all age classes exceed poor ones in average values of height and weight. This clearly better physical development of children of nobility and affluent bugeoisie was most likely related to better education and general knowledge of their parents, not just to richer diets and better living circumstances. Among nobility, 20% were owners of large landed estates who, in addition to their estate mansions, owned apartments in Warsaw where they spent a large part of the year. Many affluent children came from families of highly educated parents, such as officers and clerks (14%), physicians (10%), professional engineers (9%), and technicians (7%), teachers, (5%), industrialists (5%), lawyers, (4%), and pharmacists (3%). Among middle class, in general the most common were artisans (22%), but among the most affluent members of the middle class were army officers and clerks (11%), and lower level office workers (7%), merchants (11%), traders (9%), industrialists (7%), technical managers (4%), teachers (3%), and physicians (3%) (own calculations based on Miklaszewski 1914). Members of the working class were mostly labourers (eg. among women 50% were launderers), or petty tradespeople with most prominent among them cobblers (over 15%) and taylors (nearly 14%) (Miklaszewski 1912).

The next series of Figures (Figures 12-19) shows annual incements of body height and weight in boys and girls from privileged and impoverished groups. Data were smoothed using the LOESS method with tricube smoothing at 50% points to fit (SPSS v 22.0).

Among boys from noble and burgeois families there was present a growth spurt of 7 cm per year at age 14-15 years, while among poor boys no peak was observable. The greatest velocity among poor boys amounted to 5 cm per year that occured at age 10 years. At older ages their velocities declined systematically (Figures 12–13). Among privileged girls the peak growth velocity of 6 cm per year occurred at the age 10-11 years, while poor girls showed height increases of 4.5 cm per year in the entire age range 10-13 years without a clear peak. At the age 13-14 years poor girls achieved growth velocity of 6 cm per year, but, due to no data available for older age groups, it was impossible to ascertain whether this velocity was the largest and would not be exceeded at a later age (Figures 14-15). Analysis of velocities of body weight produced results similar to those for the height. In privileged boys there was a clear peak velocity of 5 kg per year at the age 14-15 years, while among poor boys annual velocities did not show peak velocities in the entire age range 10-19 years available for study. During this age range velocities remained moderate: 2.5 kg – 3.0 kg per year (Figures 16-17). Among privileged girls there was a clear peak of velocity – 5 kg per year – at the age

11-12 years. Among underprivileged girls at the age 11-12 years velocity was just a half of that in privileged girls. Underprivileged girls reached velocity of 5 kg per year only at the age 13-14 years following which there are no data (Figures 18-19). Growth spurt is known to be delayed in poor living conditions and this delay may be so large that the spurt will merge with the end of growth (Quetelet 1831, Dufestel 1920, Miwa 1901– za: Martin 1928, Artaria, Henneberg 2008). Since the growth spurt precedes menarche by 1-2 years, it would indicate that poor Warsaw girls would have started menstruating only at age 16 years or later.

Despite shortcomings of materials analysed here, especially the cross-sectional nature of annual velocities, it is clear that growth rates differed between social strata. Growth spurt was likely not present among the children of poor labourers. Social mobility to higher classes produced the phenomenon of growth spurt.

CONCLUSION

1. In historical populations studied here there was a significant stratification of height and body mass resulting from social stratification. Upward social mobility increased body size. In more affluent social strata, whose access to food was not limited (nobility, midle class) the significant increase in BMI has occurred already in the late 19th century producing serious prevalence of overweight and obesity. This is especially clear among newly rich enterprising middle class members, whose BMI sometimes exceeded that of nobility.
2. Membership of privileged social classes resulted in the appearance of adolescent growth spurts in height and weight. Among affluent boys peak height velocity (7 cm per year) and weight velocity (5 kg per year) occured at the age 14-15 years. Among privileged girls peak velocities (6cm and 5 kg per year) occured at the age 11-12 years. Among poor children of labourers growth spurts did not occur.

REFERENCES

Arcaleni, Emilia. 2006. "Secular trend and regional differences in the stature of Italians, 1854-1980." *Economics Human Biology* 4:24-38.

Artaria, Myrtati, and Maciej Henneberg. 2008. "The Existence of a Peak in Adolescent's Height Increments." *Folia Medica Indonesiana* 44: 196-202.

Baten, Jörg, and Matthias Blum. 2012. "An anthropometric history of the Word,1810-1980: did migration and globalization influence country trends?" *Journal of Anthropological Science* 90:1-4.

Bielicki, Tadeusz. 1981. "Niektóre antropologiczne przejawy rozwarstwienia społecznego współczesnej ludności Polski." *Kosmos* 6:563-572.

Bielicki, Tadeusz, and Zygmunt Welon. 1982. "Growth data as an indicator of social inequalities: The case of Poland." *Yearbook of Physical Anthropology* 25:153-167.

Bielicki, Tadeusz, Szczotka, Hubert, Górny, Stanisław, and Janusz Charzewski. "Rozwarstwienie społeczne współczesnej ludności Polski: analiza wysokości ciała poborowych urodzonych w 1957 r." *Przegląd Antropologiczny* 1981; 47:237-261.

Bielicki, Tadeusz, Waliszko, Anna, Hulanicka, Barbara, and Krystyna Kotlarz. 1986. "Social class gradients in menarcheal age in Poland" *Annals of Human Biology* 13:1-11.

Bielicki, Tadeusz, Welon, Zygmunt, and Wojciech Żukowski. 1988. "Problem nierówno wartości biologicznej warstw społecznych." *Materiały i Prace Antropologiczne* 109:123-140. (in Polish).

Bielicki, Tadeusz, and Anna Waliszko. 1992. "Stature, upward social mobility and the nature of statural differences between social classes." *Annals of Human Biology* 19:589-593.

Bielicki, Tadeusz, Szklarska, Alicja, Kozieł, Sławomir, and Zygmunt Welon. 2003. *Transformacja ustrojowa w Polsce w świetle antropologicznych badań 19-letnich mężczyzn.* Monografie Zakładu Antropologii PAN, nr 23. Wrocław: Zakładu Antropologii PAN. (In Polish).

Bochenek, Adam. 1906. "Materyały do charakterystyki antropologicznej ludności Królestwa Polskiego. II. Główniejsze cechy charakterystyki antropologicznej ludności włościańskiej powiatu mławskiego, guberni płockiej." *Materyały Antropologiczno-Archeologiczne i Etnograficzne* 8:69-76.

Bocheńska, Zofia. 1972. *Zmiany w rozwoju osobniczym człowieka w świetle trendów sekularnych i różnic społecznych.* Kraków: Roczniki Naukowe WSWF w Krakowie, nr 5.

Bodenhorn Howard. 2010a, "Height and body mass index values of nineteenth – century New York legislators." *Economics and Human Biology* 8(1): 121-126.

Bodenhorn, Howard. 2010b, “Height, weight and body mass index values of mid-19th century New York legislative officers.” *Economics and Human Biology* 8 (2): 291-293.

Bogin, Barry. 1999. *Patterns of human growth.* Cambridge: Cambridge University Press.

Cardoso, Hugo F. V., and Madalina Caninas. 2010. “Secular trends in social class differences of height, weight and BMI of boys from two schools in Lisbon, Portugal (1910-2000).” *Economics and Human Biology* 8: 111-120.

Carson, Scott A. 2005. “The biological standard of living in 19th century Mexico and in the American West.” *Economics and Human Biology* 3: 405-419.

Carson, Scott A. 2007. “Mexican body mass index values in the late – 19th – century American West.” *Economics and Human Biology* 5: 37-47.

Carson, Scott A. 2009. “Racial differences in body mass indices of men imprisoned in 19th Century Texas.” *Economics and Human Biology* 7: 121-127.

Carson, Scott A. 2011. “Height of female Americans in the 19th century and the antebellum puzzle.” *Economics and Human Biology* 9: 157-164.

Carson, Scott A. 2013. “Body mass, wealth, and inequality in the 19th century: Joining the debate surrounding equality and health.” *Economics and Human Biology* 11: 90-94.

Chwalba Andrzej. 2000. *Historia Polski 1795-1918.* Kraków: Wydawnictwo Literackie.

Coclanis, Peter A., and John Komlos. 1995, “Nutrition and Economic Development in Post-Reconstruction South Carolina. An Anthropometric Approach.” *Social Science History* 19: 91-115.

Cuff, Timothy. 1993. “The Body Mass Index Values of Mid-Nineteenth-Century West Point Cadets.” *Historical Methods* 26(4): 171-182.

Czapla, Zbigniew, and Grażyna Liczbińska. 2014. “Height as an indicator of economic status in the Polish territories under Prussian rule at the turn of the 19th to 20th century” *Journal of Biosocial Sciences* 46(5): 686-697.

Dufestel, Louis. 1920. *La Croissance.* Paris: Maison Gaston Doin.

Eveleth, Phyllis B., James M. Tanner. 1990. *Worldwide Variation in Human Growth.* Cambridge: Cambridge University Press.

Helmchen, Lorens A., and R. Max Henderson. 2004. “Changes in the distribution of body mass index of white US men, 1890-2000.” *Annals of Human Biology* 31(2): 174-181.

Hiermeyer, Martin. 2010. "The height and BMI values of West Point cadets after the Civil War." *Economics and Human Biology* 8: 127-133.

Jasicki, Bronislaw, Panek, Stanisław, Sikora Paweł, and Eugenia Stołyhwo. 1962. *Zarys Antropologii*. Warszawa: Państwowe Wydawnictwo Naukowe.

Kaczmarek, Maria. 1995. *Wpływ warunków życia na wzrastanie i rozwój człowieka*. Poznań : Wydawnictwo Naukowe UAM.

Kaczmarek, Maria. 2001. *Poznańskie Badania Longitudinalne. Rozwój fizyczny chłopców i dziewcząt*. Poznań: Monografie Instytutu Antropologii UAM.

Kaływa, Dobrochna. 2006. "Polska doby rozbiorów i międzywojenna," w: *Obyczaje w Polsce. Od średniowiecza do czasów współczesnych*, red. Chwalba Andrzej, Warszawa: Wydawnictwo Naukowe PWN, 221-336.

Komlos, John. 1985. "Stature and nutrition in the Habsburg Monarchy: The standard of living and economic development" *American Historical Review*. 90(5): 1149-1161.

Komlos, John. 1987. "The height and weight of West Point Cadets: dietary change in Antebellum America" *Journal of Economic History* 47(4): 897-927.

Komlos, John. 1990. "Height and Social Status in Eighteenth-Century Germany" *Journal of Interdisciplinary History* 20(4): 607-621.

Komlos, John. 2006. "The height increments and BMI values of elite Central European children and youth in the second half of the 19th century" *Annals of Human Biology* 33(3): 309-318.

Komlos, John. 2007. "Anthropometric evidence on economic growth, biological well-being and region al convergrnce In the Habsburg Monarchy, c. 1850-1910." *Journal of Historical Economics and Econometric History* 1(3): 211-237.

Kopczyński, Michał. 2006. *Wielka transformacja. Badania nad uwarstwieniem społecznym i standardem życia w Królestwie Polskim w latach 1866-1913 w świetle pomiarów antropometrycznych poborowych*, Warszawa: Oficyna Wydawnicza "Mówią wieki."

Kopczyński, Michał. 2007. "Agrarian reforms, agrarian crisis and the biological standard of living in Poland, 1844-1892." *Economics and Human Biology* 5: 458-470.

Kosieradzki K. 1905. "Przyczynek do charakterystyki fizycznej ludności męskiej powiatu Miechowskiego na zasadzie pomiarów rekrutów w ciągu lat 30-tu (1874-1903)." *Czasopismo Lekarskie* Rok VII, Tom VII: 261-281.

Kozieł, Sławomir, Szklarska, Alicja, Bielicki, Tadeusz., and Robert M. Malina. 2006. Changes in the BMI of Polish conscripts between 1965 and 2001: secular and socio-occupational variation. *International Journal of Obesity*. 30: 1382-1388.

Malina, Robert M., Bouchard, Claude, and Oded Bar-Or. 2004. *Growth, Maturation, and Physical Activity*. Champaign, Ill: Human Kinetics.

Maloney, Thomas N., and Scott A. Carson. 2008. "Living standards in Black and White: Evidence from the heights of Ohio Prison inmates, 1829-1913." *Economics and Human Biology* 6: 237-251.

Miklaszewski, Walenty. 1912. *Rozwój cielesny proletariatu Warszawy w świetle pomiarów antropometrycznych*. Warszawa: Gebethner and Wolff.

Miklaszewski, Walenty. 1914. *Rozwój cielesny klas uprzywilejowanych w Królestwie Polskiem w świetle pomiarów antropometrycznych*. Warszawa: Gebethner and Wolff.

Nicholas, Stephen, and Richard H. Steckel. 1991. "Heights and living stand arts of English workers during the early industrialization, 1770-1815." *Journal of Economic History* 51(4): 937-957.

Nowak, Oscar. 2011. *Wysokość i masa ciała młodych mężczyzn w okresie przemian historycznych i społeczno-gospodarczych drugiej połowy XIX i początku XX wieku na ziemiach polskich*. Poznań: Wydawnictwo Naukowe UAM.

Nowak, Oscar, and Janusz Piontek. 2008. "The secular trend in body height in the rural population of Podbereżce (Ukraine) in the 19th and 20th centuries." *Anthropologie* XLVI/I: 95-99.

Prus, Bolesław. 1886. "Drobne uwagi o wielkich kwestyach." *Tygodnik Ilustrowany* nr 183.

Quetelet, Adolphe. 1831. "Recherches sur la loi de croissance de l'homme." *Annales d'Hygiene Publique* 6: 89-113.

Rutkowski, Leon. 1900. "Przyczynek do antropologii krajowej." *Prawda* 25: 300-302.

Rutkowski Leon. 1906. "Charakterystyka antropologiczna ludności okolic Płońska i sąsiednich powiatów gubernii płockiej." *Materyały Antropologiczno-Archeologiczne i Etnograficzne* VIII: 3-68.

Stankiewicz, Zbigniew. 1973. "Wieś" w: *Słownik Historii Polski*, red. Łepkowski Tadeusz. Warszawa: Wiedza Powszechna, 633-654.

Staub, Kaspar, Rühli, Frank J., Woitek, Ulrich, Pfister, Christian. 2010. "BMI distribution/social stratification in Swiss conscripts from 1875 to present." *European Journal of Clinical Nutrition* 64: 335-340.

Sunder, Marco. 2004, "The height of Tennessee convicts: another piece of the 'antebellum puzzle.' " *Economics and Human Biology* 2: 75-86.

Ulijaszek, Stanley, Johnston, Francis E., and Michael A. Preece. (editors). 1998. *The Cambridge Encyclopedia of Human Growth and Development*. Cambridge: Cambridge University Press.

Walker, Robert, Gurven, Michael, Hill, Kim, Migliano, Andrea, Chagnon, Nepoleon, De Souza, Roberta, Djurovic, Gradimir et al., 2006. "Growth rates and life histories in twenty-two small-scale societies." *American Journal of Human Biology* 18: 295–311.

Welon, Zygmunt, Bielicki, Tadeusz, Jurynec Roman, Kowalczyk Janusz. 1983, "Pogłębianie się niektórych społecznych różnic w Polsce w ciągu XX stulecia, w świetle danych antropometrycznych o wzroście żołnierzy." *Kosmos* 32 (1): 5-15.

In: Biological Implications of Human Mobility ISBN: 978-1-63485-644-7
Editors: Slawomir Koziel et al.

Chapter 2

IS THE EFFECT OF AGE ON DIASTOLIC BLOOD PRESSURE MODIFIED BY THE AGE AT MIGRATION? A STUDY AMONG THE ADULT MALE SLUM DWELLERS IN EASTERN INDIA

Raja Chakraborty[1*], Slawomir Koziel[2] and Kaushik Bose[3]
[1]Department of Anthropology, Dinabandhu Mahavidyalaya, Bongaon, West Bengal
[2]Polish Academy of Sciences, Unit of Anthropology in Wroclaw
[3]Department of Anthropology, Vidyasagar University, Midnapore, West Bengal

ABSTRACT

Migration may have significant effect on both early nutrition and later life health. Migrants who come from low income families frequently settle in slum and squatter settlements, and consequently, encounter various challenges such as poor sanitation, social discrimination, low-income jobs and food scarcity. All these factors put migrants at increased health risks as compared to other subpopulations in the urban areas. On the other hand, nutrition in early childhood has its determining effect on

* Corresponding Author address: Department of Anthropology, Dinabandhu Mahavidyalaya, Bongaon, West Bengal, India. Email: rajanth2003@yahoo.co.uk.

the health status in later adult life. The objective of the present study was to examine whether the effect of migration in the early childhood had any association with the later life blood pressures among adult male slum dwellers. This cross sectional study carried out in a slum at Dum Dum, West Bengal, included 269 adult male aged 18-49 years and was born in or migrated to the slum within the age of 20 years. Current age and age at migration (AAM) were recorded for each participant. AAM were categories into three groups: Zero (born in the slum), 1-12 and 13-20 years. Height and body weight were measured to calculate body mass index (BMI). Mean blood pressures and BMI increased with age and reached the highest values in decade 40-49 years of life. Mean DBP, but not BMI, increased significantly with shift from first to second AAM group and then dropped to the lowest value in the third group. AAM had significant negative impact on DBP independent of BMI and age in both regression and covariance analyses. Thus, those who migrated at younger age, tended to have a significantly higher DBP. AAM, however, had not shown such effect on SBP independent of BMI and age. Migration to an adverse environment such as in slums during the major growth period in childhood may have adverse effect on cardio-vascular risk profile as represented by blood pressure in Bengalee males.

Keywords: blood pressure, migration, adaptation, slum dwellers

INTRODUCTION

Growth is a mirror of social circumstances and their change (Tanner 1986, Nowak 2016). The intergenerational influence (i.e., from mother to baby) of childhood nutrition suggests the necessity of having a better nutritional status in early childhood to improve the child health in the next generation (Baqui et al. 1994; Subramanian et al. 2009). Migrations from rural to urban places many times are associated with drastic change in health ecology of a population. Most of the migrants, in rural-urban migration happen to be from low income families and the majority of such migrants normally settle in slum and squatter settlements. Consequently, these migrants encounter various problems such as security, poor housing and social discrimination. All these factors put migrants at increased health risks as compared to other subpopulations in the urban areas (Afsar 2003, Ullah 2004, Roy and Abdullah 2005).

Rural-urban migration shows a significant association with overweight and obese conditions (Khan and Kraemer 2009). However, burdens of underweight and overweight were shown to coexist among lower socioeconomic groups in India (Subramanian et al. 2009). The length of U.S. residency among immigrants, especially in women, was directly associated with risk of obesity. Immigrants' risks of negative health outcomes were lower than those for U.S. born populations but increased with length of residence in the country (Barcenas et al. 2007). The risk of obesity among immigrants increased with increasing length of U.S. residency, although the risk remained significantly lower than the risk for U.S.-born individuals (Singh and Siahpush 2002; Lauderdale and Rathouz 2000).

Blood pressure is one of the markers of health conditions of adults. Elevated blood pressure (BP) is one of the most important predictors of premature mortality and morbidity due to several cardiovascular complications (Ellekjaer et al. 2001; Gu et al. 2008). BMI has been shown to be associated with BP among adults in Finland (Jousilahti et al. 1995), Sweden (Henriksson et al. 2003), Argentina (Feldstein et al. 2005), Africa (Cappuccio et al. 2008) and also in Pakistan (Khan et al. 2008). The urban poor face the worst consequences of an urbanized lifestyle, putting them at significantly high risk of non-communicable diseases (Yadav and Krishnan 2008). In a study in Kolkata, Eastern India, the prevalence of stroke incidence was greater among slum dwellers than the non-slum subjects (Das et al. 2007). Previous study with the same male population under this present study showed a positive association of BMI and age with blood pressure (Chakraborty et al. 2009). Both these parameters were at their highest mean values at the age group of 40-49, although mean height was the lowest in this age group (Chakraborty 2011). However, there is no visible study till date to delineate whether there could be some relationship between age at the time migration and blood pressure. The present study therefore attempted to demonstrate this relationship allowing for chronological age and BMI among the adult male slum dwellers.

METHODS

The Subjects and Setting

The present study was an outcome of a research project on nutritional status and health amongst the adult males residing in a slum located in the eastern part of India. The slum called Bidhan Colony was situated about 15 km off the central part of the city of Kolkata (formerly, Calcutta) in the Indian state of West Bengal. It was situated alongside the railway tracks for local trains between Dum Dum Junction and Dum Dum Cantonment railway stations under Eastern Railways. All participants belonged to Hindu caste groups speking in Bengali language, thus called *Bengalee*. Two of the three distinct parts of the slum were selected for the study through lottery. Due to operational difficulties in the field situation, no strict statistical sampling procedure was applied (Khongsdier 2002). Adult male member(s) available during the field-visits were recruited from each household. The purpose of the study was explained to participants before enrolment for the study and informed consents obtained. No household was visited more than once for recruitment of new subjects. The overall response rate was nearly 80%. Reportedly healthy men, free from any acute illness and on their normal regular work-schedule, were included. Total sample for the project comprised of 474 men aged between 18 and 84 years. It was found that 269 adult men aged 18-49 years born in or migrated to the slum within their age of 20 years. These men were selected for this present analysis.

Ethical guidelines as laid down in Helsinki declaration was adhered to (Goodyear 2007). The municipal authority and local community leaders were informed in writing before the commencement of the study.

Most participants belonged to low socio-economic background with a mean $\pm$ SD per capita income of RS 893.86 $\pm$ 507.6 (1€ ~ 75 INR). They were mostly factory workers, rickshaw-pullers and day-labourers.

Measurements

Present chronological age of the subjects and age at migration (AAM) were recorded in whole years. AAM were categories into three groups: Zero (born in the slum), 1-14 years and 15-20 years. Height and weight were recorded to the nearest 0.1 cm and 0.1 kg, respectively, using an anthropometer, and weight-scale, respectively, following standard protocol (Lohman et al. 1988). BMI was calculated as weight in kilogram (kg) divided by height in meter (m) squared. Blood pressure was measured in the right arm of each subject, using a digital blood pressure monitor with adjustable calf size, following the prescribed protocol of the manufacturer (Home Health, Switzerland). The accuracy of the device was routinely checked during the entire study period. Resting systolic (SBP) and diastolic (DBP) blood pressures (in mmHg) were recorded with the subject in a sitting position after at least 15 minutes of rest prior to measurement and again, for a second time, at least 5 minutes after the first reading. The mean value of two readings was finally recorded (Chakraborty et al. 2011).

Statistical analysis

The significance of differences in SBP, DBP and BMI across age groups and AAM groups were assessed by one way ANOVA. The effect of duration spent in the slum after migration on blood pressure was assessed by multiple regression and analysis of covariance (ANCOVA). The systolic and diastolic blood pressures were dependent variables in separate analyses. The chronological age and age at migration (AAM) were independent factors/ predictors, whereas the BMI was covariate. The covariate model was implemented as a Generalised Linear Model (GLM) and the value of Wald's chi-square statistic was used to evaluate the significance of the effect of independent variables as well as the covariate and also the second order interaction between independent variable and the covariate. This result were also presented graphically. All calculations were performed using Statistica 10.0 (StatSoft 2011).

RESULTS

Table 1 shows the numbers and percentage of males by three groups of chronological age and AAM. Table 2 presents descriptive statistics of blood pressures and BMI by chronological age and AAM categories, respectively. Mean DBP increased with age and reached the highest values in decade 40-49 years of life ($F = 15.85$, $df = 2$, $p < 0.001$). SBP however did not increase significantly ($F = 2.806$, $df = 2$, $p = 0.062$). Similar increase was also noticed for BMI ($F = 5.723$, $df = 2$, $p < 0.005$). There was small change in systolic blood pressure between two first groups of AAM, but in group of 13-20 years, the mean dropped down. The change however was not significant. There was however a significant curvilinear change in the mean values of diastolic blood pressure across the three groups of AAM. Mean DBP increased from first to second group, reaching the highest value and then drops to lowest value at the highest AAM group ($F = 5.736$, $df = 2$, $p < 0.005$). Little change (not significant) was noticed in mean of BMI across the groups.

Table 1. Number (%) of persons according to chronological age and age at migration

Age groups (years)	Age at migration (years)		
	Born	1-12 years	13-20 years
18-29	104 (72.22)	23 (26.14)	16 (41.03)
30-39	35 (24.31)	31 (35.23)	16 (41.03)
40-49	5 (3.47)	34 (38.64)	13 (33.33)

Table 2. Means, standard deviations of blood pressure and BMI by categories of age and age at migration in slum dwellers

	N	SBP	DBP	BMI
Age (years)				
18-29	142	117.6 (12.6)	76.6 (8.2)	20.0 (3.09)
30-39	76	119.0 (24.31)	80.6 (7.7)	20.9 (3.15)
40-49	51	122.8 (18.5)	84.5 (12.7)	21.7 (3.98)
AAM (years)				
Born	143	119.0 (10.9)	78.5 (8.2)	20.5 (3.0)
1-12	88	119.6 (15.6)	81.8 (7.7)	20.9 (3.8)
13-20	38	117.4 (17.5)	76.1 (9.6)	20.2 (3.3)

Table 3. Linear regression analysis of BP on age, age at migration and BMI

DV IV	B	SeB	Beta	T	p
DBP					
Constant	51.99	3.50	---	14.87	0.0001
BMI	0.91	0.16	0.31	5.72	0.0001
Age	0.38	0.07	0.35	5.82	0.0001
AAM	-1.91	0.77	0.14	-2.48	0.0014
SBP					
Constant	87.94	5.31	---	16.55	0.0001
BMI	1.38	0.24	0.33	5.70	0.0001
Age	0.15	0.10	0.10	1.52	0.129
AAM	-1.21	1.17	-0.06	-1.03	0.303

Table 3 presents the results of multiple regression analyses of diastolic BP on age, AAM (as continuous) and BMI. Age at migration had significant negative effect ($t = -2.48$, $p < 0.05$) on DBP independent of BMI and age. That is to say, those who migrated at a younger age, tended to have a significantly higher DBP. AAM, however, had not shown such effect on SBP independent of BMI and age.

The results of analysis of covariance (ANCOVA) are presented in Table 4. AAM had significant effect ($F = 5.50$; $p < 0.01$) on DBP, after allowing for age and BMI. However, BMI and age also had highly significant independent effects on DBP. A second order interaction between age and AAM also highly significantly influenced the values of DBP. No such effect of AAM was evident when another ANCOVA was run for SBP as dependent factor (results not shown). Weighted means with 95% CI of diastolic BP of male slum dwellers are shown on Figure 1, which shows that among men in age group 40-49, those who migrated during 1-12 years of age, had the highest DBP.

Table 4. Results of analyses of covariance where DBP was dependent and age and age at migration (coded) were independent variables with BMI as covariate

Variable	Sum of Squares (df)	Mean square	F	p
Intercept	16911.639 (1)	16911.639	271.603	0.0001
BMI	1422.967 (1)	1422.967	22.853	0.0001
Age	4151.106 (31)	133.907	2.151	0.001
AAM	684.883(2)	342.441	5.500	0.005
Age*AAM	4426.045(38)	116.475	1.871	0.003

*Interaction between age and AAM (coded).

DISCUSSION

The present study, although in a small scale, demonstrated that the age at the time of migration to a slum had significant statistical effect on diastolic blood pressure and this effect was independent of their chronological age and BMI. The effect was negative, which indicated that the earlier was the age at migration, the higher was the probability of having higher DBP in adulthood, particularly, during 40-49 years of age. It was observed that both SBP and DBP and also BMI increased from the lowest to the highest age group. When AAM was used as a categorical variable, the individuals who migrated during their 1-12 years of age had shown noticeably the highest mean DBP. In regression analysis with all variables being continuous, AAM showed significant independent effect on DBP but not on SBP. When AAM as a categorical variable was used in ANCOVA, with its three categories, it was observed that this variable also had highly significant independent effect on DBP but not on SBP. More interestingly, the interaction between age and AAM was also significant. It implied that AAM might have differential effect in different age groups or vice versa.

Previous studies on the same group of men with larger sample size showed that BMI had significant impact on blood pressure, particularly on DBP (Chakraborty et al. 2009, Chakraborty 2011). The men also had the highest body weight and BMI in age group 40-49 years (Chakraborty 2011). Moreover this age group also had significantly the highest mean DBP. However, this study demonstrated that the adult men at this age group demonstrated differential DBP profile given their different age at migration.

Childhood place of residence was shown to be one of the strong predictors for the health related outcomes (Khan and Kraemer 2008). There were many studies showing relationship of body size and mass with the length of residence in a new place after migration. The first generation immigrants had a lower body size than second- and third generation immigrants, and that among the immigrants the length of residence in the U.S. was associated with increased body size (Gordon-Larsen et al. 2003; Sundquist and Winkleby 2000; Popkin and Udry 1998; Kaplan et al. 2004,). Singh and Siahpush (2002) found that immigrants' risks of negative health outcomes, including obesity, were lower than those for U.S. born populations but increased with length of residence in the country. Kaplan and colleagues (2004) also found that recent Hispanic immigrants are generally healthier than the U.S.-born population, and that this distinction tends to diminish over time as immigrants adapt to a different sociocultural environment. Goel et al. (2004) found that length of residence in the U.S. is associated with higher BMI among immigrants resident in the U.S. for 10 years or more, and that obesity rates among immigrants who have been in the country for at least 15 years approach those of native born U.S. residents. While Lauderdale and Rathouz (2000) found a low prevalence of overweight among Asian Americans, they also found the prevalence of overweight to be higher for those with longer length of residence in the U.S.

Our present study indicated that the age at migration could also have a long term negative health effect like higher blood pressure. The men who migrated with their parents at the age between 1 and 12 years had higher mean DBP in comparison with those who born at the slum or those who migrated into the slum at a later age. Many among the slum dwelling adult males under this present study had migrated to this slum in various stages of their life. Interviews with a number of participants of this study revealed that the migration to this slum were accompanied with a number of adversities and associated with unhealthy conditions and poverty. Age between 1 and 12 years is somewhat a period of major growth and development. The mean age at peak height velocity of the Bengalee boys was reported to be aroud 13 yeras (de Onis et al. 2001). The poverty and struggle to survive by the newly migrated parents of the children aged 1-12 years might have resulted nutritional stress. Early life events like nutritional insult may have adverse effect on later life health. Early growth has a 'critical window' period. Environmental stimulus in this period may determine later life health outcome to a significant extent (Wells 2003). The programming of metabolic load vs. capacity in early growth period may also have later life health implications (Wells 2007, 2013). The relative economic improvement and adopting modern obesogenic lifestyle including consumption of calorie rich food by these set of children in their adulthood might have increased body mass followed by increasing blood pressure. A study on these same slum dwelling men showed that a small improvement in income could lead to a much larger increase in soft drink consumption (Chakraborty et al. 2009a). It was suggested that gaining body weight might be a reflection of relatively higher socioeconomic status than a migrated group had in their country of origin, especially if they were unaware of the potential health hazards associated with gaining weight (Barcenas et al. 2007).

We propose that the the impact of adverse environment as a consequence of economic struggle of the family resulting in nutritional and other stresses during the major growth period preceding adolescence might have adverse effect on adult blood pressures in slum dwelling Bengalee males. The study has its limitations of small sample size and lack of direct evidence of early growth period environment etc. However further in depth and focused study on this aspect may establish the independent effect of age at the time of migration on adverse adult cardiovascular profile (e.g., blood pressure) in a later life.

REFERENCES

Afsar, Rita. 2003. Internal migration and the development nexus: the case of Bangladesh. Dhaka: Refugee and Migratory Movements Research Unit,

Accessed March 15, 2016. http://www.migrationdrc.org/publications/working_papers/WP-C2

Barcenas, Barcenas H., Wilkinson, Anna V., Strom, Sara S., Cao, Yumei, Saunders, Katherine C., Mahabir, Somdat, Herna´ndez-Valero, Marı´a A., Forman, Michele R., Spitz, Margaret R. and Melissa L. Bondy. 2007. "Birthplace, Years of Residence in the United States, and Obesity Among Mexican-American Adults." *Obesity* 15: 1043-52.

Baqui, Abdullah H., Arifeen, Shams E., Amin, Samir, and Robert E. Black. 1994. "Levels and correlates of maternal nutritional status in urban Bangladesh." *European Journal of Clinical Nutrition* 48:349-57.

Cappuccio, Francesco P., Kerry, Sally M., Adeyemo, Adebowale, Luke, Amy, Amoah, Albert G.B., Bovet, Pascal, Connor, Myles D., Forrester, Terrence, Gervasoni, Jean-Pierre, Kaki, Gisela K, Plange-Rhule, Jacob, Thorogood, Margaret, and Richard S. Cooper. 2008. "Body size and blood pressure: an analysis of Africans and the African diaspora." *Epidemiology* 19: 38-46.

Chakraborty, Raja, Bose, Kaushik and Samiran Bisai. 2009. "Body mass index and blood pressure among adult Bengalee male slum dwellers of Kolkata, India." *Journal of Public Health* 17:301-08.

Chakraborty, Raja. 2011. *Nutritional Status of Bengalee Adult Males in Bidhan Colony: An Urban Slum in Dum Dum, West Bengal.* Ph.D. Thesis (Unpublished). Vidyasagar University, MIdnapore, West Bengal, India.

Das, Shyamal K., Banerjee, Tapas K., Biswas, Atanu, Roy, Trishit, Raut, Deepak K., Mukherjee, Chandra S., Chaudhuri, Arijit, Hazra, Avijit, and Jayanta Roy. 2007. "A prospective community based study of stroke in Kolkata, India." *Stroke* 38: 906–910.

De Onis, Mercedes, Dasgupta, Parasmani, Saha, Shyamal, and Sengupata, Debasish, and Monika Blössner. 2001. "The national centre for health statistics reference and the growth of Indian adolescent boys." *American Journal of Clinical Nutrition* 74:238-53.

Ellekjaer, Hanne, Holmen, Jostein, and Lars Vatten. 2001. "Blood pressure, smoking and body mass in relation to mortality from stroke and coronary heart disease in the elderly. A 10-year follow-up in Norway." *Blood Press*ure 10: 156–163

Feldstein, Barcenas A., Akopian, Maia, Olivieri, Antonio O., Kramer, Agustina P., Nasi, Magdalena, and Delia Garrido. 2005. "A comparison of body mass index and waist-to-hip ratio as indicators of hypertension risk in an urban Argentine population: a hospital-based study." *Nutrition Metabolism and Cardiovascular Disease* 15: 310-315.

Goel, Mita S., McCarthy Ellen P., Phillips, Russel S., and Christina C Wee. 2004. "Obesity among US immigrant subgroups by duration of residence." *Journal of American Medical Association* 292(23):2860-2867

Goodyear, Michael D. E., Krleza-Jeric, Karmela, and Trudo Lemmens. 2007. "The declaration of helsinki." *British Medical Journal* 335:624-625.

Gordon-Larsen, Penny, Harris Kathleen M., Ward, Dianne S., and Barry M. Popkin. 2003. "Acculturation and overweight-related behaviors among Hispanic immigrants to the US: the National Longitudinal Study of Adolescent Health." *Social Science and Medicine* 57(11):2023-34.

Gu, Qiuping, Burt, Vicki L., Paulose-Ram, Ryne, Yoon, Sarah, Richard F. Gillum. 2008. "High blood pressure and cardiovascular disease mortality risk among U.S. adults: the third National Health and Nutrition Examination Survey mortality follow-up study." *Annals of Epidemiology* 18: 302–309.

Henriksson, Karin M., Lindblad, Ulf, Gullberg, Bo, Ågren, Bo, Nilsson-Ehle, Peter, and Lennart Råstam. 2003. "Body composition, ethnicity and alcohol consumption as determinants for the development of blood pressure in a birth cohort of young middle-aged men." *European Journal of Epidemiology* 18: 955–963.

Jousilahti, Pekka, Tuomilehto, Jaakko, Vartiainen, Erkki, Valle, Timo, and Aulikki Nissinen. 1995. "Body mass index, blood pressure, diabetes and the risk of antihypertensive drug treatment: 12-year follow-up of middle-aged people in eastern Finland." *Journal of Human Hypertension* 9: 847–854

Kaplan, Mark S., Huguet, Nathalie, Newsom, Jason T, and Bentson H. McFarland. 2004. "The association between length of residence and obesity among Hispanic immigrants." *American Journal of Preventive Medicine* 27: 323-326

Khan, Mobarak M. H., and Alexander Kraemer. 2009. "Factors associated with being underweight, overweight and obese among ever-married non-pregnant urban women in Bangladesh." *Singapore Medical Journal* 50: 804-813.

Khan, Asmatullah, Haq, Faheem U., Pervez, Mohammad B., Saleheen, Danish, Frossard, Philippe M., Ishaq, Mohammad, Hakeem, Abdul, Sheikh, Hamza T., and Usma Ahmad. 2008. Letter to the editor: "Anthropometric correlates of blood pressure in normotensive Pakistani subjects." International Journal of Cardiology 29:259–262.

Khongsdier, Ramendra. 2002. "Body mass index and morbidity in adult males of the War Khasi in Northeast India." *European Journal of Clinical Nutrition* 56:484-89.

Lauderdale, Diane S., and Paul J. Rathouz. 2000. "Body mass index in a US national sample of Asian Americans: effects of nativity, years since immigration and socioeconomic status." *International Journal of Obesity and Related Metabolic Disorders* 24:1188 –94.

Lohman, Timothy G., Roche, Alex F., and Reynaldo Martorell. 1998. Anthropometric Standardization Reference Manual. Chicago: Human Kinetics Books.

Nowak-Szczepanska, Natalia, Gomula, Aleksandra, Ipsen, Marie J., and Slawomir Koziel. 2016. Different effects of living conditions on the variation in BMI and height in children before the onset of puberty. *European Journal of Clinical Nutrition* doi: 10.1038/ejcn.2016.30.

Popkin Barry M., and J. Richard Udry. 1998. "Adolescent obesity increases significantly in second and third generation U.S. immigrants: the National Longitudinal Study of Adolescent Health." *Journal of Nutrition* 128(4):701-706.

Roy, Gouri S., and A. Q. M. Abdullah. 2005. "Assessing needs and scopes of upgrading urban squatters in Bangladesh." *BRAC University Journal* 2:33-41.

Singh, Gopal K., and Mohammad Siahpush. 2002. Ethnic-immigrant differentials in health behaviors, morbidity, and cause-specific mortality in the United States: an analysis of two national databases. *Human Biology* 74:83–109.

StatSoft Inc. 2010. *STATISTICA Version 10.0*. www.statsoft.com.Statsoft Inc., Tulsa, USA.

Subramanian, Sv V., Perkins Jessica M., Kashif T. Khan. 2009. Do burdens of underweight and overweight coexist among lower socioeconomic groups in India? *American Journal of Clinical Nutrition* 90:369-76.

Sundquist, Jan and Marilyn Winkleby. 2000. "Country of birth, acculturation status and abdominal obesity in a national sample of Mexican-American women and men." *International Journal of Epidemiology* 29(3):470-477.

Ullah, A. K. M. Ahsan. 2004. "Bright city lights and slums of Dhaka city: determinants of rural-urban migration in Bangladesh." Migration Letters 1:26-41.

Wells, Jonathan C. K. 2003. "The thrifty phenotype hypothesis: thrifty offspring or thrifty mother?" *Journal of Theoretical Biology* 221:143-161.

Wells, Jonathan C. K. 2007. "The programming effect in early growth." *The Early Human Development* 83: 743-58.

Wells, Jonathan C. K. 2013. "The thrifty phenotype: an adaptation in growth or metabolism?" *American Journal of Human Biology* 23:65-75.

Yadav, Kapil, Krishnan, Anand. 2008. "Changing pattern of diet, physical activity and obesity among urban, rural and slum populations in North India." *Obesity Reviews* 9:400–408.

In: Biological Implications of Human Mobility
Editors: Slawomir Koziel et al.
ISBN: 978-1-63485-644-7

Chapter 3

SOCIOECONOMIC STATUS, BLOOD PRESSURE AND HYPERTENSION

Jyoti Ratan Ghosh*
Department of Anthropology, Visva Bharati University,
West Bengal, India

ABSTRACT

Hypertension is an important cause of mortality and morbidity. Epidemiological studies have revealed that hypertension is a 'silent killer,' with no signs of warning. However, little is known about how socioeconomic status is related to blood pressure and hypertension in India. Thus, the present study examined the association of socioeconomic status with blood pressure and hypertension. A total of 254 adult males were incorporated in the present cross-sectional study. Data on occupation, education and monthly income were collected and utilized as measures of socioeconomic status. Systolic blood pressure (SBP) and diastolic bold pressure (DBP) were taken and hypertension was defined as SBP $\geq$ 140 mm/Hg and/or DBP $\geq$ 90 mm/Hg. One way analysis of variance, chi-square and regression analysis were used to understand the association of socioeconomic status with blood pressures and hypertension.The prevalence of hypertensionin the studied population was 17.7%. The result revealed that both SBP and DBP were higher among individuals with lower level of education andwho belongs to

* Corresponding Author address: Department of Anthropology, Visva-Bharati, Santiniketan-731235, West Bengal, India. Email: jrghosh@rediffmail.com; jrghosh@visva-bharati.ac.in.

lower occupational groups.There were also significant differences in the prevalence of hypertension according to different educational and occupational groups. However, no significant association was observed between income and blood pressure. Both education and occupation were inversely associated with hypertension and explained 11.3% and 10.9% of variance of hypertension, respectively. Thus, the present study demonstrated that socioeconomic status in terms of education and occupation but not by income was strong predictor of blood pressure progression and hypertension in adult male.

Keywords: blood pressure, developing countries, education, hypertension, India, income, occupation, socioeconomic status

INTRODUCTION

Hypertension is an important cause of mortality and morbidity (Gulliford et al. 2004). Hypertension accounted for 9.4 million deaths and 7% of disability adjusted life years in 2010 (Lim et al. 2013). The overall prevalence of hypertension in India is 29.8% (Anchala et al. 2014) and it is directly responsible for 57% of all stroke deaths and 24% of all coronary heart disease (CHD) deaths in India (Gupta 2004).The global trends indicating that hypertension is increasing rapidly in lower and middle income countries (Mohan et al. 2013).

Socioeconomic status covers a wide range of measures including education, income, occupation, living conditions, and many other socioeconomic aspects of life (Kaplan and Keil, 1993). An association between socioeconomic status and health has been long been known. Studies in developed countries demonstrated that socioeconomic status was inversely associated with blood pressure and hypertension (Jaglal and Goel, 1994, Colhoun et al. 1998, Cois and Ehrlich, 2014). However, little is known about how socioeconomic variables influence the distribution of blood pressure and hypertension in developing countries, because, studies from developing countries demonstrated heterogeneous relation between socioeconomic status and blood pressure including positive, negative and no association (Mendez et al. 2003, Gulliford et al. 2004). Thus, a better understanding of the reason for this heterogeneity in association is needed for population based prevention strategies. Since hypertension is one of the most important established risk factors for coronary heart disease, type 2 diabetes mellitus and renal disease (Ghosh and Bandyopadhyay 2007, Conen et al. 2009, Kinra et al. 2014). Study showed that a small difference in blood pressure between socio-economic groups could have a substantial impact on cardiovascular disease mortality, because a 5 mm/Hg lower population mean diastolic blood pressure predicts a 21% lower coronary heart disease rate and a 34% lower stroke rate (Colhoun et al. 1998).

Although socioeconomic differences in blood pressure and hypertension are described from many countries, very few studies have been done in India, especially in eastern India (Gilberts et al. 1994, Kinra et al. 2014). Thus, using education, occupation and income as socioeconomic indicators, this paper examines the association of socioeconomic status with blood pressures and hypertension.

Material and Methods

The present study was conducted in sub-urban areas of West Bengal, India. A total of 254 adult males within the age range of 18-69 years were randomly recruited to participate in the present cross-sectional study. Informed consent was obtained from all participants before the work commenced. Individuals under medication were excluded. All participants were asked to complete a questionnaire that includes specific information on age, occupation, education, monthly income and medical history. Self-reported occupation, education and monthly income were utilized as measures of socioeconomic status (Salonen, 1982, Lynch et al. 1996, Mendez et al. 2003, Padyab and Norberg 2014). Education, occupation and income were used because these were the most relevant indicator socioeconomic status to understand the differences in health status (Galobardes et al. 2006, Kim et al. 2014). Occupation was categorized in to two groups, higher occupational groups (including service and business) and lower occupational groups (labourer, van puller etc.). Educational status was categorized into four groups according to the years of education (group I: <5 years, group II: 5 years - <10 years, group III: 10 years - <15 years and group IV: ≥15 years). Income quartiles were used to categorized monthly income into four groups (group I: ≤ Rs. 3000.00, group II: Rs. 3001.00 to ≤6000.00, group III: Rs.6001.00 to ≤10000.00 and group IV: ≥Rs. 10001.00). However, the highest three categories of income and education were combined during interpretation of the result.

SBP and DBP measurements were taken following standard procedure (Weiner and Lourie 1981). Hypertension (SBP ≥ 140 mm/Hg and/ or DBP ≥

90 mm/Hg) was defined according to the Seventh Report of The Joint National Committee (JNC 7) recommendation (Chobanian et al. 2003).

Descriptive statistics were computed by mean, standard deviation (SD) and frequency distribution. Pearson correlation coefficient was used to understand the relationship of age with SBP and DBP. Analysis of variance (ANOVA) was used to test the effect of occupation, education and monthly incomeon SBP and DBP. Chi-square test was used to understand the association between socioeconomic status and hypertension. Regression analyses were used to understand the relationship of socioeconomic measures with blood pressure and hypertension. Statistical analysis was performed using the Statistical Package for Social Sciences version 9.0.0. A p-value of <0.05 was considered as significant.

RESULTS

Characteristics of the studied population are presented in Table 1. It shows that majorities (35.4%) of the studied participants have more than equal to fifteen years of education, followed by 31.5% participants with ten years to less than fifteen years of education. However, 15.7% and 17.3% participants have less than five years and five to less than equal to ten years of education, respectively. Regarding occupation, 25.2%of the participants belong to lower occupational group and rests (74.8%) were in higher occupational group. With regard to the income, 28.3% participant's income was Rs. 3000.00 or below, followed by 25.2% with income between Rs. 3001.00 to ≤6000.00 and 26% participants with income between Rs.6001.00 and ≤10000.00. However, 20.5% participants income was Rs. 10001.00 or above. Regarding hypertension, its prevalence in the studied population was 17.7%.

The mean age of the studied population was 43.55 (10.24) years. The mean SBP and DBP was 121.26 (15.00) mm/Hg and 77.3 (10.63) mm/Hg, respectively (Table 2). Results of the Pearson correlation analysis revealed that age was not significantly ($p>0.05$) correlated with both SBP and DBP.

Socioeconomic status wise distributions of blood pressures are presented in Table 3. It revealed that both SBP and DBP were comparatively higher among individuals with lower level of education (<5 years) and there were significant ($p < 0.01$) differences in both SBP and DBP among different educational status groups. The result also revealed that individuals who belongs to lower occupational groups had significantly ($p < 0.01$) higher SBP and DBP (Figure 1). However, there were no significant (p>0.05) differences in blood pressure (SBP and DBP) among different income groups.

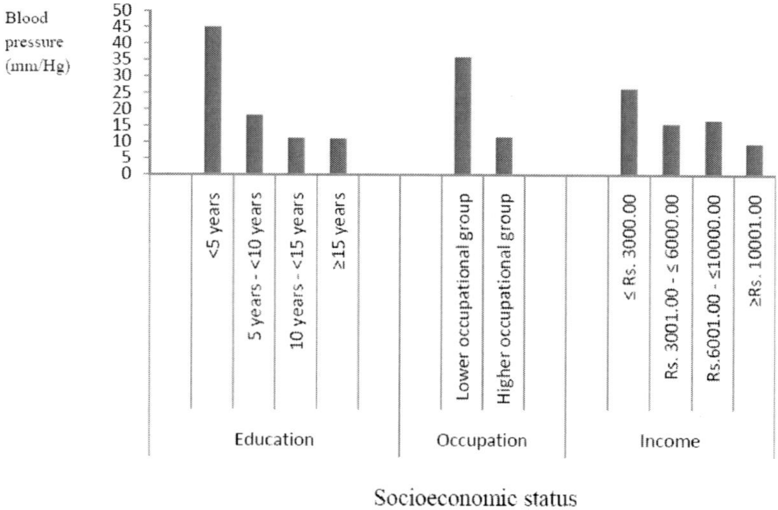

Figure 1. Socioeconomic status and blood pressures.

Table 1. Characteristics of the studied population

Variables	Characteristics	N	%
Education			
	<5 years	40	15.7
	5 years -<10 years	44	17.3
	10 years - <15 years	80	31.5
	≥15 years	90	35.4
Occupation			
	Lower occupational group	64	25.2
	Higher occupational group	190	74.8
Income			
	≤Rs. 3000.00	72	28.3
	Rs. 3001.00 - ≤ 6000.00	64	25.2
	Rs.6001.00 - ≤10000.00	66	26.0
	≥Rs. 10001.00	52	20.5
Hypertension status			
	Normal	209	82.3
	Hypertension	45	17.7

Table 2. Mean and standard deviation of age and blood pressures

Variables	Mean	SD
Age (year)	43.55	10.24
Systolic blood pressure (mm/Hg)	121.26	15.00
Diastolic blood pressure (mm/Hg)	77.3	10.63

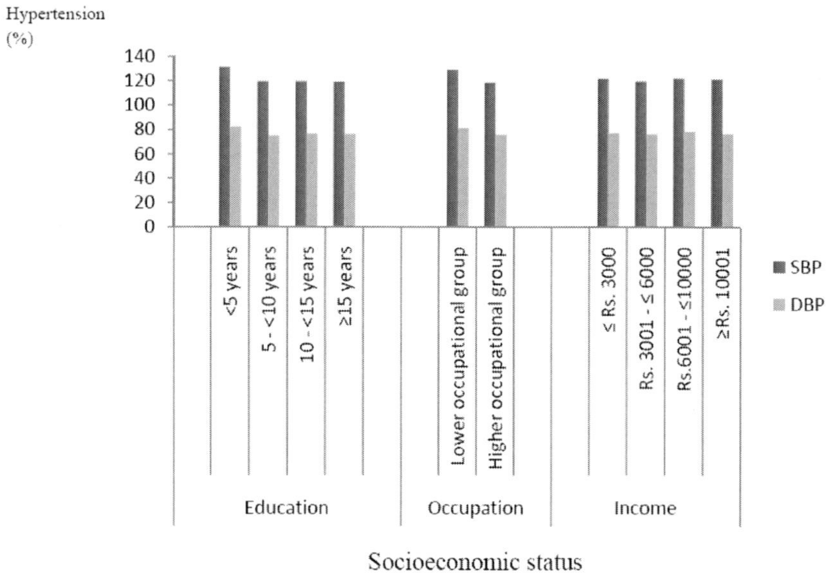

Figure 2. Socioeconomic status and hypertension.

The linear regression analysis (Table 4) of socioeconomic variables with blood pressures revealed that educational status was the significant predictor of both SBP ($p < 0.01$) and DBP ($p < 0.05$), but the association was inverse (SBP b = -3.11; DBP b = -1.30). Moreover, educational status explained 4.9% and 1.7% of the variances of SBP and DBP, respectively. Like education, occupational status was also inversely (SBP b = -10.47; DBP b = -5.42) associated and significant ($p < 0.01$) predictors of both SBP and DBP. Occupational status explained 9.2% and 4.9% of the variances of SBP and DBP, respectively. However, income status was not the significant ($p>0.05$) predictors of SBP and DBP.

Table 3. Socioeconomic status wise distribution of blood pressures

Variables	Characteristics	SBP		F	*p*	DBP		F	*p*
Education		Mean	SD			Mean	SD		
	<5 years	131.20	15.57	7.48	<0.01	82.3	11.66	3.85	<0.01
	5-<10 years	119.51	15.10			75.22	10.74		
	10 - <15 years	119.53	15.67			76.8	11.89		
	≥15 years	119.25	12.37			76.55	8.14		
Occupation									
	Lower occupational group	129.09	14.92	25.56	<0.01	81.36	11.09	13.05	<0.01
	Higher occupational group	118.63	14.12			75.94	10.14		
Income									
	≤ Rs. 3000	121.84	17.87	0.35	>0.05	77.08	11.70	0.46	>0.05
	Rs. 3001 - ≤ 6000	119.64	12.31			76.61	9.71		
	Rs.6001 - ≤10000	122.08	16.07			78.6	12.05		
	≥Rs. 10001	121.43	12.28			76.83	8.06		

Table 4. Linear regression analysis of socioeconomic variables with blood pressures

Dependent variables	Independent variables	b	R	R^2	F	*p*
Systolic blood pressure						
	Education	-3.11	0.221	0.049	12.99	<0.01
	Occupation	-10.47	0.303	0.092	25.56	<0.01
	Income	0.10	0.007	0.00005	0.01	>0.05
Diastolic blood pressure						
	Education	-1.30	0.131	0.017	4.37	<0.05
	Occupation	-5.42	0.222	0.049	13.05	<0.01
	Income	0.17	0.018	0.0003	0.08	>0.05

Table 5. Socioeconomic status wise distribution of hypertension

Variables	Characteristics	N	%	χ^2	*p*	Nagelkerke R^2	*p*
Education	<5 years	18	45.00	25.42	<0.01	0.113	<0.01
	5 years - <10 years	8	18.18				
	10 years - <15 years	9	11.25				
	≥15 years	10	11.11				
Occupation	Lower occupational group	23	35.94	19.48	<0.01	0.109	<0.01
	Higher occupational group	22	11.58				
Income	≤Rs. 3000.00	19	26.39	6.30	<0.05	0.034	<0.01
	Rs. 3001.00 - ≤ 6000.00	10	15.62				
	Rs.6001.00 - ≤10000.00	11	16.67				
	≥Rs. 10001.00	5	9.61				

Socioeconomic status wise distributions of hypertension in the studied population (Table 5) revealed significant ($\chi^2 = 25.42$, $p < 0.01$) differences in the prevalence of hypertension among different educational status groups (Figure 2). Similar significant ($\chi^2 = 19.48$, $p < 0.01$) differences in the prevalence of hypertension was also observed between occupational groups. But there was no significant ($\chi^2 = 6.30$, $p>0.05$) association between hypertension and income status. Moreover, the result also revealed that educational status explained 11.3% (Nagelkerke $R^2 = 0.113$, $p < 0.01$) of variance of hypertension. However, occupational and income status explained 10.9% (Nagelkerke $R^2 = 0.109$, $p < 0.01$) and 3.4% (Nagelkerke $R^2 = 0.034$, $p < 0.01$) of variance of hypertension, respectively.

DISCUSSION

The result of the present study to understand the association of socioeconomic status with blood pressures and hypertension revealed significantly higher mean blood pressure among individuals with lower level of education and who belongs to lower occupational group. But, there was no consistent association between blood pressures and income. Not only blood pressures, but the prevalence of hypertension was also higher among individuals with lower level of education and occupation. However, the relationship between blood pressure and socioeconomic status was independent of age. The results also demonstrated that education and occupation explained about 11% of variance of hypertension. In accordance with the present study, Conen et al. (2009) also observed that education but not income was associated with BP progression and incident hypertension. A previous study to understand the relationship between grade of employment and coronary risk factors also demonstrated that men in lower employment grades had higher blood pressure than men in the higher grades (Marmot et al. 1978). Kanjilal et al. (2006) also observed higher prevalence of hypertension among adults with less education. James et al. (2006) showed that a lower socioeconomic position was associated with a double risk of hypertension. However, the lack of significant association of income with blood pressure and hypertension may be due to the fact that income fluctuates over time and does not necessarily reflect an individual's wealth, which may influence individual's health behavior (Conen et al. 2009).

In contrary to the present study some previous studies in South Indian demonstrated higher blood pressures and hypertension in the highest socioeconomic group compared to the lowest socioeconomic group (Gilberts et al. 1994, Reddy et al. 2002). Other studies also observed similar inconsistencies in the relationships of socioeconomic status with blood pressures in developing countries. For example, some studies revealed negative associations, some revealed positive associations and even some studies also observed no association between socioeconomic status and blood pressure variables (Colhoun et al. 1998, Gulliford et al. 2004). The basis for this inconsistency in associations between socioeconomic status and blood pressure in developing countries was unclear. However, one possible reason may be the ethnic variation and heterogeneity in stages of modernization and economic development (Colhoun et al. 1998, Mendez et al. 2003).

Studies also demonstrated that the inverse association of socioeconomic positions with hypertension may be partly responsible for the higher incidence of cardiovascular disease among individuals with lower socioeconomic status (Kanjilal et al. 2006, Conen et al. 2009). Life course studies demonstrated that low childhood socioeconomic position was also associated with higher risk of hypertension in adulthood (James et al. 2006, Gebreab et al. 2015). Framingham offspring study (Loucks et al. 2010) and the 1946 British birth cohort study (Jones et al. 2015) showed that not only the traditional cardiovascular risk factors like hypertension, but novel biological risk markers of coronary heart disease like intercellular adhesion molecule-1, tumor necrosis factor II, lipoprotein phospholipase A2, monocyte chemoattractant protein-1, fibrinogen, C-reactive protein, proinsulin, interleukin-6 and leptin were also inversely associated with life course socioeconomic position.

The inverse relationship of socioeconomic status with increasing blood pressures and hypertension may be related to the socioeconomic differences in health related behaviors like smoking, alcohol consumption and inadequate nutrition, which are more prevalent in lower social classes (Kaplan and Keil 1993). Study (Colhoun et al. 1998) also demonstrated that the stress of unemployment or job insecurity in lower socioeconomic status may also be related to the higher blood pressures. Access and quality of care as well as disadvantaged neighborhood environment also mediates the inverse relationship between socioeconomic status and hypertension (Trivedi et al. 2006, Cozier et al. 2007, Conen et al. 2009). Moreover, higher prevalence of hypertension in lower socioeconomic status may be associated with reduced awareness, treatment, and control of hypertension (Mendez et al. 2003).

In conclusion, the present cross-sectional study demonstrated that socioeconomic status in terms of education and occupation was significantly and negatively associated with blood pressure and hypertension. Moreover, in this population, education and occupation was stronger predictor of hypertension than income. Thus this study in developing country demonstrated a trend of relationship which was commonly observed in developed countries. However, the main limitation of the present study was the relatively small sample, and therefore was not representative of the Indian population. Because of vast ethnic and cultural heterogeneity among Indians, it is important to study other ethnic groups to see if the trends observed also exist among them.

The results obtained from such studies could be used to prevent hypertension and cardiovascular risk.

Acknowledgments

The author is grateful to all participants. The author is also grateful to the Department of Anthropology, Visva-Bharati for providing all the facilities for conducting the research.

References

Anchala, Raghupathy, Kannuri, Nanda K., Pant, Hira, Khan, Hassan, Franco, Oscar H., Di Angelantonio, Emanuele, and Dorairaj Prabhakaran. 2014. "Hypertension in India: a systematic review and meta-analysis of prevalence, awareness, and control of hypertension." *Journal of Hypertension* 32:1170–1177.

Chobanian, Aram V., Bakris, George L., Black, Henry R., Cushman, William C., Green, Lee A., Izzo, Joseph L. Jr., Jones, Daniel W., Materson, Barry J., Oparil, Suzanne, Wright, Jackson T. Jr, Roccella, Edward J., and The National High Blood Pressure Education Program Coordinating Committee. 2003. "The seventh report of the joint national committee on prevention, detection, evaluation, and treatment of high blood pressure: the JNC 7 report." *Journal of American Medical Association* 289:2560-2572.

Cois, Annibale, and Rodney Ehrlich. 2014. "Analyzing the socioeconomic determinants of hypertension in South Africa: a structural equation modeling approach." *BMC Public Health* 14:414.

Colhoun, Helen M., Hemingway, Harry, and Neil R. Poulter. 1998. "Socio-economic status and blood pressure: an overview analysis." *Journal of Human Hypertension* 12: 91–110.

Conen, David, Glynn, Robert J., Ridker, Paul M., Buring, Julie E., Albert, Michael A,. 2009. "Socioeconomic status, blood pressure progression, and incident hypertension in a prospective cohort of female health professionals." *European Heart Journal* 30:1378–1384.

Cozier, Yvette C., Palmer, Julie R., Horton, Nicholas J., Fredman, Lisa., Wise, Lauren A., and Lynn Rosenberg. 2007. "Relation between neighborhood median housing value and hypertension risk among black women in the United States." *American Journal of Public Health* 97:718–724.

Galobardes, Bruna., Shaw, Mary, Lawlor, Debbie A., Lynch, John W., George D. Smith. 2006. "Indicators of socioeconomic position (part 1)." *Journal of Epidemiology and Community Health* 60:7-12.

Gebreab Samson Y., Roux, Ana V. D., Brenner, Allison B., Hickson, DeMare A., Sims, Maria, Subramanyam, Malavika, Griswold, Michael E., Wyatt, Sharon B., and Sherman A. James. 2015. "The impact of life course socioeconomic position on cardiovascular disease events in African Americans: The Jackson heart study." *Journal of American Heart Association* 4:e001553.

Ghosh, Jyoti R., and Arup R Bandyopadhyay. 2007. "Comparative evaluation of obesity measures: relationship with blood pressures and hypertension." *Singapore Medical Journal* 48:232-235.

Gilberts, Ericus C., Arnold, Mark J., and Diederick E. Grobbee. 1994. "Hypertension and determinants of blood pressure with special reference to socioeconomic status in a rural south Indian community." *Journal of Epidemiology and Community Health* 48:258-261.

Gulliford, Martin C., Mahabir, Deepak, and Brian Rocke. 2004. "Socioeconomic inequality in blood pressure and its determinants: cross-sectional data from Trinidad and Tobago." *Journal of Human Hypertension* 18:61–70.

Gupta, Rajiv. "Trends in hypertension epidemiology in India." *Journal of Human Hypertension* 18:73-78.

Jaglal, Susan B., and Vivek Goel. 1994. "Social inequity in risk of coronary artery disease in Ontario." *Canadian Journal of Cardiology* 10:439–443.

James, Sherman A., Hoewyk, John V., Belli, Robert F., Strogatz, David S., Williams David R., Trevillore E. Raghunathan. 2006. "Life-course socioeconomic position and hypertension in African American men: The Pitt county study." *American Journal of Public Health* 96:812–817.

Jones, Rebecca, Hardy, Rebecca, Sattar, Naveed, Deanfield, John E., Hughes, Alun, Kuh Diana, Murray, Emily T., Whincup, Peter H., Claudia Thomas. 2015. "Novel coronary heart disease risk factors at 60-64 years and life course socioeconomic position: The 1946 British birth cohort." *Atherosclerosis* 238:70-76.

Kanjilal, Sanjat, Gregg Edward W., Cheng, Yiling J., Zhang, Ping, Nelson, David E., Mensah, George, and Gloria L. A. Beckles. 2006. "Socioeconomic status and trends in disparities in 4 major risk factors for cardiovascular disease among US adults, 1971–2002." *Archives of Internal Medicine* 166: 2348–2355.

Kaplan George A., and Julian E. Keil. 1993. "Socioeconomic factors and cardiovascular disease: A review of the literature." *Circulation* 88:1973–1998.

Kim, Jeong H., Jeong, Myung, H., Park, In H., Choi, Jin S., Rhee, Jung A., Lee, Doo H., Park, Soo H., Kim, In S., Jeong Hae C., Cho, Jae Y., Jang, Jang Y., Lee, Ki H., Park, Keun-Ho, Sim, Doo S., Kim, Kye H., Hong, Young J., Park, Hyung W., Kim, Ju H., Ahn Youngkeun, Cho, Jeong G., and Jong C. Park. 2014. "The Association of socioeconomic status with thee-year clinical outcomes in patients with acute myocardial infarction who underwent percutaneous coronary intervention." *Journal of Korean Medical Science* 29:536-543.

Kinra, Sanjay, Johnson, Mark, Kulkarni, Bhushan, Rameswar Sarma, K. V., Ben-Shlomo, Yoah, and Davey G. Smith. 2014. "Socio-economic position and cardiovascular risk in rural indian adolescents: evidence from the Andhra Pradesh children and parents study (APCAPS)." *Public Health* 128:852-859.

Lim, Stephen S., Vos, Theo, Flaxman, Abraham D., Danaei, Goodarz, Shibuya, Kenji, Adair Rohani, Heather, et al. 2013. "A comparative risk assessment of burden of disease and injury attributable to 67 risk factors and risk factor clusters in 21 regions, 1990-2010: a systematic analysis for the Global Burden of Disease Study 2010." *Lancet* 380:2224-2260.

Loucks, Eric B., Pilote, Louise., Lynch, John W., Richard, Hugues., Almeida, Nisha D., Benjamin, Emelia J., and Joanne M. Murabito. 2010. "Life course socioeconomic position is associated with inflammatory markers: the Framingham offspring study." *Social Science Medicine* 71:187-195.

Lynch, John W., Kaplan, George A., Cohen, Richard D., Tuomilehto, Jaakko, and Jukka T. Salonen. 1996. "Do cardiovascular risk factors explain the relation between socioeconomic status, risk of all-cause mortality, cardiovascular mortality, and acute myocardial infarction?" *American Journal of Epidemiology* 144:934-942.

Marmot, M. G., Adelstein, A. M., Robinsonn, N., Rose, G. A. 1978. "Changing social-class distribution of heart disease." *British Medical Journal* 2:1109-1112.

Marmot Michael G., Rose Geoffrey, Shipley Martin, and Patric J.S. Hamilton. 1978. "Employment grade and coronary heart disease in British civil servants." *Journal of Epidemiology and Community Health* 32:244-249.

Mendez, Michelle A., Cooper, Richard, Wilks, Rainford, Luke, Amy, and Terrence Forrester. 2003. "Income, education, and blood pressure in adults in Jamaica, a middle-income developing country." *International Journal of Epidemiology* 32:400–408.

Mohan, Sailesh, Campbell, Norm, Arun Chockalingam. 2013. "Time to effectively address hypertension in India." *Indian Journal of Medical Research* 137:627-631.

Padyab, Mojgan, and Margaretta Norberg. 2014. "Socioeconomic inequalities and body mass index in Vasterbotten County, Sweden: a longitudinal study of life course influences over two decades." *International Journal of Equity in Health* 13:35.

Reddy, Kanala K. R., Rao, Alahari P., Thavanati P. K. Reddy. 2002. "Socioeconomic status and the prevalence of coronary heart disease risk factors." *Asia Pacific Journal of Clinical Nutrition* 11:98–103.

Salonen, Jukka T. 1982. "Socioeconomic status and risk of cancer, cerebral stroke, and death due to coronary heart disease and any disease: a longitudinal study in eastern Finland." *Journal of Epidemiology and Community Health* 36:294-297.

Trivedi, Amal N., Zaslavsky, Alan M., Schneider, Eric C., and John Z Ayanian. 2006. Relationship between quality of care and racial disparities in Medicare health plans. *Journal of the American Medical Association* 296:1998–2004.

Weiner Joseph S., and John A. Lourie. 1981. *Practical Human Biology*. New York: Academic Press.

In: Biological Implications of Human Mobility ISBN: 978-1-63485-644-7
Editors: Slawomir Koziel et al.

Chapter 4

HIGHER GRADE OF SOCIOECONOMIC STATUS IS ASSOCIATED WITH THE INCREASED PHYSICAL GROWTH OF SCHOOLCHILDREN OF KOLKATA, INDIA

Satabdi Ghosh, Tusharkanti Ghosh* and Ananga Mohan Chandra
Department of Physiology, University of Calcutta, India

ABSTRACT

Age dependent physical growth of the school children is influenced by environmental factors such as socio-economic status (SES) of the family in addition to genetic factors. The effect of vertical social mobility of a family in a population on the physical characters of its members may be indirectly investigated by measuring the SES of the families along with physical parameters (such as height and weight) of the children in a cross section of that population. This study was designed to find out the relationship between the SES of the family and age dependent growth of height and weight of school children (age: 5-12 years) of Kolkata. Height and weight were assessed by standard techniques and BMI was calculated from height and weight. Stunting and underweight of 624 school children (304 boys and 320 girls) were assessed from height-for-age (HAZ) and

* Corresponding Author address: Neurophysiology Laboratory, Department of Physiology, University of Calcutta, 92, A.P.C. Road, Kolkata Email: tusharkantighosh53@yahoo.in.

weight-for-age (WAZ) Z scores respectively with reference values of World Health Organization. SES was measured from updated Kuppuswamy scale. A large portion of boys belong to lower middle and upper lower SES whereas a large portion of girls belong to lower middle and upper middle SES. The overall prevalence of stunting and underweight was 20.35% (boys: 25.33%, girls: 15.63%) and 15.54% (boys: 20.39%, girls: 10.94%) respectively. The SES of the school children was significantly correlated with height, HAZ, weight and WAZ ($p < 0.001$). Height and weight were strongly associated with their SES, age and sex ($p < 0.001$). Age related growth of height and weight are influenced by the SES of the family.

Keywords: height, weight, social mobility, socio-economic status

INTRODUCTION

Social mobility can be defined as the movement of individuals or groups in social positions over time (Boundless 2015). It may be of two types, vertical and horizontal. Vertical mobility occurs in stratified societies and involves a shift of a person's position on attained education or wealth or power or prestige, whereas, horizontal mobility involves movement of individual to a different place of residence. Education, occupation and income are important factors influencing social mobility (Nunn et al. 2007). Therefore the socio-economic status (SES) is important index of the social mobility. Poor SES may affect physical growth like height and weight by selection of unhealthy diet, unhealthy behaviors and by affecting physical activity (Pampel et al. 2010). Poor SES may also affect mental health, psychological well-being and cognitive and emotional development of children (Adler and Rehkopf 2008; Bradley and Corwyn 2002; McLoyd 1998; Conger and Donnellan 2007).

In a stratified society, inequality is a part of the social structure which may lead to inequalities in child growth (Hancock et al. 2015). SES is measured on the basis of education, income and occupation of the members of the family (Kumar et al. 2012; Conger et al. 2010). The effect of vertical social mobility on the physical growth may be measured indirectly in a cross-sectional study as it would indicate the possible changes of the physical parameters in the event of vertical mobility of the family from the same community. Kolkata is a major metropolitan city in India and the inhabitants of this city belong to different socio-economic classes. The influence of SES of the family on the physical parameters of the Kolkata children has not been studied in recent years. Therefore, the present study has been undertaken to find out the effect of different classes of SES on the height and weight of Kolkata children (5-12 years of age) which may indicate indirectly the influence of vertical social mobility of families of school children of Kolkata.

Methods

Study Population

The study was conducted on 624 school children (304 boys and 320 girls) aged 5–12 years selected randomly from six primary and secondary schools of five different zones (east, west, north, south and central) of Kolkata metropolitan area of West Bengal. At least one school was randomly selected from each zone. Name and date of birth of the students were obtained from the register of schools after getting permission from the school authorities and parents to carry out the work. The protocol and procedures employed in this study were in accordance with the Ethical Guidelines for biomedical research on human subjects of the Indian Council of Medical Research (2006). Children with any systemic disease or children who have undergone major surgical operations or children with neurological, sensory or motor problems were excluded from this study.

Nutritional Status

Their nutritional statuses were calculated from height-for-age (HAZ) and weight-for-age (WAZ) Z scores by using the reference values of WHO (World Health Organization 1983). Children with Z scores between -1 and -1.99 of both indices were considered as mild undernourished, children with Z scores between -2 and -2.99 of both indices were considered as moderate undernourished and children with Z scores below -3 of both indices were considered as severely undernourished.

Socio-Economic Status

Socio-economic status (SES) was measured using the updated Kuppusswami scale (Kumar et al. 2012), which is based on parental education, parental occupation and monthly family income.

Statistical Analyses

Mean and standard error of mean were computed for each anthropometric variable according to age and sex. ANOVAs were conducted to test for differences in the parameters by sex. Pearson's correlation coefficient was used to evaluate the relationship between the anthropometric variables and SES. Stepwise linear regression models were used to find out the association between dependent (e.g., height and weight) and independent variables (e.g., SES, age and sex). Chi-square test was performed to examine the association between nutritional status and SES. The statistical analyses were performed using SPSS (Version 20.0, Armonk, NY).

RESULTS

SES: The distribution of subjects to different socio-economic statuses is presented in Table 1. Maximum number of boys belong to upper lower (35.86%) and lower middle (33.52%) SES whereas maximum number of girls belong to the lower middle (38.75%) and upper middle (35.63%) SES.

Height and weight: The mean height and weight of school children are presented in Table 2. Height and weight were increased with the advancement of age in both sexes.

Nutritional status from HAZ: Table 3 showed the distribution of subjects to different grades of nutritional status (stunting as per HAZ). The percentages of boys belong to mild, moderate and severe undernourished groups were 26.32%, 19.74% and 5.59% respectively. Whereas 21.88%, 11.88% and 3.75% girls belong to mild, moderate and severe undernourished groups respectively.

Nutritional status from WAZ: Table 4 showed the distribution of subjects to different grades of nutritional status (underweight as per WAZ). The percentages of boys belong to mild and moderate undernourished groups were 32.89% and 20.39% respectively. On the other hand, 27.50% and 10.94% girls belong to the mild and moderate undernourished groups respectively. But no one was found in the severe undernourished category as per WAZ.

Table 1. Distribution of subjects (in numbers and percentages) to different socio-economic status

Socio-economic status (SES)	Boys [n_1=304] (%)	Girls [n_2=320] (%)	Total [N=624] (%)
Upper	20 (6.58)	22 (6.88)	42 (6.73)
Upper middle	58 (19.08)	114 (35.63)	172 (27.56)
Lower middle	102 (33.52)	124 (38.75)	226 (36.22)
Upper Lower	109 (35.86)	52 (16.25)	161 (25.80)
Lower	15 (4.93)	8 (2.50)	23 (3.69)

n_1=number of boys, n_2=number of girls, N= (n_1+ n_2) = total number of subjects.

Table 2. Mean and standard error of mean values of height and weight of school children of Kolkata

Age (years) [n_1, n_2]	Boys (n_1 = 413)		Girls (n_2 = 432)	
	Height (cm)	Weight (Kg)	Height (cm)	Weight (Kg)
5 [39, 41]	105.43 ± 1.20	18.10 ± 0.57	107.17 ± 1.12	18.32 ± 0.62
6 [38, 36]	114.58 ± 0.94	20.32 ± 0.47	113.66 ± 1.07	19.81 ± 0.92
7 [37, 54]	120.49 ± 0.76	21.32 ± 0.60	119.56 ± 0.89	21.89 ± 0.65
8 [34, 33]	126.60 ± 1.10	25.40 ± 1.00	123.85 ± 1.19	25.56 ± 1.35
9 [37, 38]	129.15 ± 1.15	26.62 ± 1.03	128.48 ± 1.28	26.83 ± 1.16
10 [38, 38]	130.41 ± 1.20	27.45 ± 1.10	137.16 ± 1.00***	35.54 ± 1.57***
11 [39, 41]	137.99 ± 1.23	31.15 ± 1.08	143.61 ± 1.18**	40.07 ± 1.52***
12 [42, 39]	141.43 ± 1.09	34.25 ± 0.96	150.71 ± 1.57***	47.04 ± 1.71***

Values = Mean ± standard error of mean.

n_1 = number of school boys, n_2 = number of school girls.

**$p < 0.01$, ***$p < 0.001$ compared to the same age group of boys.

Table 3. Distribution of subjects (in numbers and percentages) to different grades of HAZ

Grades of nutritional status from HAZ	Boys [n_1=304] (%)	Girls [n_2=320] (%)	Total [N=624] (%)
Normal (+1 to -0.99)	147 (48.36)	200 (62.50)	347 (55.61)
Mild (-1 to -1.99)	80 (26.32)	70 (21.88)	150 (24.04)
Moderate (-2 to -2.99)	60 (19.74)	38 (11.88)	98 (15.71)
Severe (< -3)	17 (5.59)	12 (3.75)	29 (4.65)
Total stunting (moderate+ severe)	77 (25.33)	50 (15.63)	127 (20.35)

n_1=number of boys, n_2=number of girls, N= (n_1+ n_2) = total number of subjects.

Table 4. Distribution of subjects (in numbers and percentages) to different grades of WAZ

Grades of nutritional status from WAZ	Boys [$n_1 = 304$] (%)	Girls [$n_2 = 320$] (%)	Total [N = 624] (%)
Normal (+1 to -0.99)	142 (46.71)	197 (61.56)	339 (54.33)
Mild (-1 to -1.99)	100 (32.89)	88 (27.50)	188 (30.13)
Moderate (-2 to -2.99)	62 (20.39)	35 (10.94)	97 (15.54)
Severe (< -3)	0 (0.00)	0 (0.00)	0 (0.00)
Total underweight (moderate + severe)	62 (20.39)	35 (10.94)	97 (15.54)

n_1=number of boys, n_2=number of girls, N= (n_1+ n_2) = total number of subjects.

Table 5. Pearson correlation between SES and height, weight, HAZ, WAZ of school children of Kolkata

		Height	**Weight**	**HAZ**	**WAZ**
SES	**Pearson Correlation**	0.308***	0.321***	0.572***	0.457***

*** p < 0.001.

Table 6. Stepwise linear regression models of the association between height and SES, age

Independent variable	Beta 95% confidence interval		p value	R^2	Changes in R^2 from previous model
	Unstandardized	Standardized			
Model 1					
SES	4.638	0.308	0.000***	0.095	0.095
Model 2					
SES	4.287	0.285	0.000***	0.821	0.726
Age	5.380	0.853	0.000***		

Dependent variable: Height, ***p < 0.001.

Pearson correlation between different parameters: Table 5 showed the Pearson correlation between SES and height, weight, HAZ, WAZ. SES of the surveyed children was significantly correlated with the above mentioned parameters ($p < 0.001$).

Association between physical parameters (height and weight) and SES, age, sex: Association between height and SES, age, sex is shown in Table 6. Height was strongly associated with SES and age. Table 7 showed the association between weight and SES, age, sex. Weight was strongly associated with SES, age and sex.

Table 7. Stepwise linear regression models of the association between height and SES, age and sex

Independent variable	Beta 95% confidence interval		p value	R^2	Changes in R^2 from previous model
	Unstandardized	Standardized			
Model 1					
SES	3.444	0.321	0.000***	0.103	0.103
Model 2					
SES	3.240	0.302	0.000***	0.584	0.481
Age	3.122	0.694	0.000***		
Model 3					
SES	2.955	0.275	0.000***	0.598	0.014
Age	3.138	0.697	0.000***		
Sex	2.573	0.123	0.000***		

Dependent variable: Weight, ***$p < 0.001$.

Table 8. Distribution of subjects to grades of nutritional status from HAZ according to SES

Socio-economic status (SES)	Grades of nutritional status from HAZ				Total
	Normal (%)	Mild (%)	Moderate (%)	Severe (%)	
Boys (n = 304)					
Upper	19 (95.00)	1 (5.00)	0 (0.00)	0 (0.00)	20 (100.00)
Upper middle	54 (93.10)	3 (5.17)	1 (1.72)	0 (0.00)	58 (100.00)
Lower middle	48 (47.06)	35 (34.32)	18 (17.65)	1 (0.98)	102 (100.00)
Upper lower	23 (21.10)	37 (33.95)	36 (33.03)	13 (11.93)	109 (100.00)
Lower	3 (20.00)	4 (26.67)	5 (33.33)	3 (20.00)	15 (100.00)
$\chi^2 = 115.711$, df = 12, $p < 0.001$					
Girls (n = 320)					
Upper	21 (95.45)	1 (4.55)	0 (0.00)	0 (0.00)	22 (100.00)
Upper middle	100 (87.72)	10 (8.77)	2 (1.75)	2 (1.75)	114 (100.00)
Lower middle	71 (57.26)	38 (30.65)	15 (12.10)	0 (0.00)	124 (100.00)
Upper lower	8 (15.38)	19 (36.54)	20 (38.46)	5 (9.62)	52 (100.00)
Lower	0 (0.00)	2 (25.00)	1 (12.50)	5 (62.50)	8 (100.00)
$\chi^2 = 261.100$, df = 12, $p < 0.001$					
Total (n = 624)					
Upper	40 (95.24)	2 (4.76)	0 (0.00)	0 (0.00)	42 (100.00)
Upper middle	154 (89.53)	13 (7.56)	3 (1.74)	2 (1.16)	172 (100.00)
Lower middle	119 (52.65)	73 (32.30)	33 (14.60)	1 (0.44)	226 (100.00)
Upper lower	31 (19.25)	56 (34.78)	56 (34.78)	18 (11.18)	161 (100.00)
Lower	3 (13.04)	6 (26.09)	6 (26.09)	8 (34.78)	23 (100.00)
$\chi^2 = 359.038$, df = 12, $p < 0.001$					

Relationship between HAZ and SES: A significant association was found between the distribution of children in different grades of of nutritional status from HAZ (normal, mild, moderate and severe) and different classes of SES (boys: $\chi 2 = 115.711$, df = 12, $p < 0.001$; girls: $\chi 2 = 261.100$, df= 12, $p < 0.001$; total: $\chi 2 = 359.038$, df = 12, $p < 0.001$) (Table 8). A large percentage of children having lower classes of SES were present in the mild, moderate and severe undernourished groups.

Relationship between WAZ and SES: A significant association was found between the distribution of children in different grades of of nutritional status from WAZ (normal, mild and moderate) and different classes of SES (boys: $\chi 2 = 31.605$, df = 8, $p < 0.001$; girls: $\chi 2 = 127.919$, df = 8, $p < 0.001$; total: $\chi 2 = 143.663$, df = 8, $p < 0.001$) (Table 9). A large percentage of children having lower classes of SES were present in the mild and moderate undernourished groups.

Table 9. Distribution of subjects to grades of nutritional status from WAZ according to SES

Socio-economic status (SES)	Grades of nutritional status from WAZ				Total
	Normal (%)	Mild (%)	Moderate (%)	Severe (%)	
Boys (n = 304)					
Upper	17 (85.00)	2 (10.00)	1 (5.00)	0 (0.00)	20 (100.00)
Upper middle	48 (82.76)	9 (15.52)	1 (1.72)	0 (0.00)	58 (100.00)
Lower middle	39 (38.24)	38 (37.25)	25 (24.51)	0 (0.00)	102 (100.00)
Upper lower	34 (31.19)	45 (41.28)	30 (27.52)	0 (0.00)	109 (100.00)
Lower	4 (26.67)	6 (40.00)	5 (33.33)	0 (0.00)	15 (100.00)
$\chi^2 = 31.605$, df = 8, $p<0.001$					
Girls (n = 320)					
Upper	19 (86.36)	3 (37.50)	0 (0.00)	0 (0.00)	22 (100.00)
Upper middle	97 (85.09)	15 (13.16)	2 (1.75)	0 (0.00)	114 (100.00)
Lower middle	70 (56.45)	41 (33.06)	13 (10.48)	0 (0.00)	124 (100.00)
Upper lower	11 (21.14)	26 (50.00)	15 (28.85)	0 (0.00)	52 (100.00)
Lower	0 (0.00)	3 (13.64)	5 (62.50)	0 (0.00)	8 (100.00)
$\chi^2 = 127.919$, df = 8, $p < 0.001$					
Total (n = 624)					
Upper	36 (85.71)	5 (11.90)	1 (2.38)	0 (0.00)	42 (100.00)
Upper middle	145 (84.30)	24 (13.95)	3 (1.74)	0 (0.00)	172 (100.00)
Lower middle	109 (48.23)	79 (34.96)	38 (16.81)	0 (0.00)	226 (100.00)
Upper lower	45 (27.95)	71 (44.10)	45 (27.95)	0 (0.00)	161 (100.00)
Lower	4 (17.39)	9 (39.13)	10 (43.48)	0 (0.00)	23 (100.00)
$\chi^2 = 143.663$, df = 8, $p < 0.001$					

DISCUSSION

The subjects of the present study belonged to different grades of SES (upper, upper middle, lower middle, upper lower and lower) and it was found from the results that subjects having higher classes of SES had better physical growth than that of the lower classes of SES. The height and weight of the surveyed school children were increased with the advancement of age. But the age dependent growth of height and weight were dependent on the SES of their families. Regression analyses suggest that height and weight of surveyed school children were strongly associated with their SES. These findings indicate if the children of the present study having lower classes of SES move to higher SES, their physical health will be better than that of the present physical health and vice versa. Therefore, the effect of vertical social mobility on the physical growth has been indicated from this cross-sectional study.

SES may affect physical growth by influencing different factors such as quality of nutrition, health care etc. (Babar et al. 2010; Green-Hernandez et al. 2001; Hackman et al. 2010). As physical growth is closely related with the food availability to the children (linked with the economic status of the family) and awareness of nutrients contents of food of their parents, the SES is likely to influence the age dependent growth of physical parameters. In the present study the nutritional status of the surveyed children was assessed from the HAZ and WAZ. An individual's height-for-age helps to identify a child who is stunted or short in height due to prolonged or long term undernutrition (World Health Organization 2008). Whereas, weight-for-age reflects body weight relative to the child's age on a given day (World Health Organization 2008). Weight-for-age is an indicator of underweight and is influenced by both the height of the child (height-for-age) and his or her weight (weight-for-height or wasting). Therefore, nutritional status as per WAZ indicates both prolonged and acute nutritional status. The large portion of children having higher classes of SES (upper, upper middle) were present in the normal nourished groups whereas a large portion of children having lower classes of SES (lower middle, upper lower and lower) were present in the mild, moderate and severe chronic undernourished groups. Similarly a large portion of children having higher classes of SES (upper, upper middle) were present in the normal nourished groups as per WAZ, whereas a large portion of children having lower classes of SES (lower middle, upper lower and lower) were present in the mild and moderate underweight groups. Similar findings were reported in school children (8.00–10.99 years aged) of Guatemalan (Groeneveld et al. 2007). Higher prevalence of stunting in lower SES groups than that of middle and higher SES has also been reported by several workers (Mushtaq et al. 2011; Van de Poel et al. 2008). It was also found that the prevalence of underweight was higher among children having lower SES (Bharati et al. 2010).

SES may affect physical activity as more space is available in house in higher SES, and there is greater chance of physical activity of the children (Tandon et al. 2012; Saccani et al. 2013; Pampel et al. 2010). Moreover, children from higher SES get better tools and opportunity for physical activity (Saccani et al. 2013). Health care provided to the children of higher SES are also better than that of lower SES (Becker and Newsom 2003). Psychological stress is an important factor for physical growth. The children are likely to get an environment of psychological well-being in families of higher SES (Adler and Rehkopf 2008; Bradley and Corwyn 2002; McLoyd 1998; Conger and Donnellan 2007). Therefore, many factors are potential influencing components of SES and several authors have considered SES as an independent factor for the development of physical growth (Ali 2013; Brito and Noble 2014; Tomalski et al. 2013).

This study indicates that SES has a strong dynamic influence on age related growth of height and weight in school children of Kolkata and these physical parameters will probably be influenced in the event of vertical social mobility of the family of the children as SES is one of the important factors determining the social mobility of any family.

ACKNOWLEDGMENTS

We would like to acknowledge the authorities of primary and secondary schools (Akra High Madrasa for Boys, Khidirpore Academy, Deshbandhu Vidyapith, Kendriya Vidyalaya, Bethune Collegiate School and Ramjoy Seal Sishu Pathshala) for providing permission to carry out the work with special mention to Krishnendu Sarkar, Prasun Haldar, Rituja Ghosh, Subhadeep Dutta Gupta, Mrityunjay Sarkar and also to Md. Waris, Shankar Bandopadhyay, Dr. Kuntal Biswas, Sampita Das for their continuous help during this work.

REFERENCES

Adler, Nancy E., and David H. Rehkopf. 2008. "U.S. disparities in health: descriptions, causes, and mechanisms." *Annual Review of Public Health* 29:235–252.

Ali, Syed S., 2013. "A brief review of risk-factors for growth and developmental delay among preschool children in developing countries". *Advanced Biomedical Research* 2:91.

Babar, Nabeela F, Muzaffar, Rizwana, Khan, Muhammad A., and Seema Imdad. 2010. "Impact of Socioeconomic Factors on Nutritional Status in Primary School Children." *Journal of Ayub Medical College, Abbottabad* 22:15–18.

Becker, Gay and Edwina Newsom. 2003. "Socioeconomic status and dissatisfaction with health care among chronically ill African Americans." *American Journal of Public Health* 93:742–748.

Bharati, Susmita, Chakrabarty, Suman, Som, Suparna, Pal, Manaranjan and Premananda Bharati. 2010. "Socio–economic determinants of underweight children in West Bengal, India." *Asian Pacific Journal of Tropical Medicine* 3:322–327.

Boundless 2015. *Types of Social Mobility.* Accessed October 3, 2015. https://www.boundless.com/sociology/textbooks/boundless-sociology-textbook/stratification-inequality-and-social-class-in-the-u-s-9/social-mobility-76/types-of-social-mobility-452-2421/.

Bradley, Robert H., and Robert F. Corwyn. 2002. "Socioeconomic Status and Child Development." *Annual Review of Psychology* 53:371–399.

Brito, Natalie H., and Kimberley G. Noble. 2014. "Socioeconomic status and structural brain development." *Frontiers in Neuroscience* 8:276.

Conger, Rand D., Conger, Katherine J., and Monica Martin. 2010. "Socioeconomic Status, family processes, and individual development." *Journal of Marriage and the Family* 72:685–704.

Conger, Rand D., and M. Bennet Donnellan. 2007. "An interactionist perspective on the socioeconomic context of human development." *Annual Review of Psychology* 58:175–199.

Green-Hernandez, C., Singleton, J.K. and Aronzon, D.Z., 2001. "Healthy Growth and Development of the School-Age Child." In *Primary Care Pediatrics.* Edited by Raymond C. Baker, 141–150. Philadelphia, PA: Lippincott Williams and Wilkins.

Groeneveld, Iris F., Solomons, Noel W., and Colleen M. Doak. 2007. "Nutritional status of urban schoolchildren of high and low socioeconomic status in Quetzaltenango, Guatemala." *Pan American Journal of Public Health* 22:169–177.

Hackman, Daniel A., Farah, Martha J., and Michael J. Meaney. 2010. "Socioeconomic status and the brain: mechanistic insights from human and animal research." *Nature Reviews Neuroscience* 11:651–659.

Hancock, Caroline, Bettiol, Silvana, and Lesley Smith. 2015. "Socioeconomic variation in height: analysis of National Child Measurement Programme data for England." *Archives of Disease in Childhood.*

Kumar, Neeta, Gupta, Neeru, and Jugal Kishore. 2012. "Kuppuswamy's socioeconomic scale: updating income ranges for the year 2012." *Indian Journal of Public Health* 56:103–104.

McLoyd, Vonnie C. 1998. "Socioeconomic disadvantage and child development." *The American Psychologist* 53:185–204.

Mushtaq, Mohammad U., Gull, Sibgha, Khurshid, Usman, Shahid, Ubeera, Shad, Mushtaq Ahmad, and Arif Mahmood Siddiqui. 2011. "Prevalence and socio-demographic correlates of stunting and thinness among Pakistani primary school children." *BMC Public Health* 11:790.

Nunn, Alex, Johnson, Steve, Monro, Surya, Bickerstaffe, Tim, and Sarah Kelsey. 2007. *Factors Influencing Social Mobility.* Norwich: Corporate Document Services.

Pampel, Fred C., Krueger, Patrick M., and Justin T. Denney. 2010. "Socioeconomic Disparities in Health Behaviors." *Annual Review of Sociology* 36:349–370.

Saccani, Raquel, Valentini, Nadia C., Pereira, Kelia R. G., Muller, Alessandra B., and Carl Gabbard. 2013. "Associations of biological factors and affordances in the home with infant motor development." *Pediatrics International* 55:197–203.

Tandon, Pooja S., Zhou, Chuan, Sallis, James F., Cain, Kelli L, Frank, Lawrence D., and Brian E. Saelens. 2012. "Home environment relationships with children's physical activity, sedentary time, and screen time by socioeconomic status." *The International Journal of Behavioral Nutrition and Physical Pctivity* 9:88.

Tomalski, Przemyslaw, Derek G. Moore, Ribeiro, Helena, Axelsson, Emma, and Elizabeth Murphy. 2013. "Socioeconomic status and functional brain development - associations in early infancy." *Developmental Science*, 16:676–687.

Van de Poel, Ellen, Hosseinpoor, Ahmad R., Speybroeck, Niko, Ourti, Tom Van, and Jeanette Vega. 2008. "Socioeconomic inequality in malnutrition in developing countries." *Bulletin of the World Health Organization* 86:282–291.

World Health Organization, 1983. *Measuring change in nutritional status : guidelines for assessing the nutritional impact of supplementary feeding programmes for vulnerable groups,* 64–100. Geneva: World Health Organisation.

World Health Organization, 2008. *Training Course on Child Growth Assessment*, Geneva: World Health Organisation.

In: Biological Implications of Human Mobility ISBN: 978-1-63485-644-7
Editors: Slawomir Koziel et al.

Chapter 5

DIFFERENCES IN GROWTH AND DEVELOPMENT BETWEEN MOSCOW SCHOOLCHILDREN DUE TO THEIR FAMILY STATUS

Elena Godina*, Irena Khomyakova and Ludmila Zadorojnaya
Institute and Museum of Anthropology, Moscow State University, Moscow, Russia

ABSTRACT

The study is based on cross-sectional data of Russian children and adolescents from 7 to 17 (1422 boys and 1530 girls) collected in Moscow in 1996-2000. They were attending different types of schools - municipal schools of ordinary type and new, "elite" schools: lyceums, gymnasia, etc. The program included about 50 anthropometrical measurements Questionnaire contained information on parental profession and education, family income, number of children in the family, birth order, birth weight and length of the investigated individual. Three categories of parental education - separately for fathers and mothers - were used: 1 – school education (at least 8 classes); 2 – specialized secondary education; 3 – higher education. Gender differences in the reaction to these variables

* Corresponding Author address: Institute and Museum of Anthropology, Moscow State University, Moscow, Russia, Email: godina@antropos.msu.ru.

were revealed. In boys most of the characteristics were increasing parallel to father's education. In girls the biggest values of many measurements, particularly those relevant to body weight and fat component, were shown in group 1. Chest circumference and chest index (chest shape) were also bigger in the daughters of less educated fathers. There were no statistically significant differences for stature in girls of those three groups. For mothers' education the direction of differences in boys is practically the same as for father's education. Girls seem not responsive at all to the influence of this independent variable. According to number of children per family, the direction of differences is the same: in both sexes body dimensions are decreasing parallel with increasing number of children. However, in girls the variations are more pronounced: 0.3 to –0.7 S v. 0.2 to -0.3 S in boys. Morphology of children from schools of different type was also compared with very similar conclusions: boys from municipal schools show much lower values in most measurements, while girls show lower body and leg length but higher body weight, trunk length, chest circumference and fat layer, particularly on trunk (centripetal).

Keywords: Growth, development, schoolchildren, family status

INTRODUCTION

The end of the last century in Russia was a period of considerable political and economic changes. In terms of human biology it was shown that stabilization or stagnation of growth in length took place in the 1990's (Godina 1998, Godina 2001, Sukharev and Mikhailova 2000, Yampolskaya 2000). Thus, mean values for stature of 17-year old boys and girls from Moscow have not changed since early 1980's. The tempos of sexual maturation also did not change much remaining practically at the same level, as could be seen from the mean age of development of secondary sexual characteristics and menarcheal age. At the same time a strong trend towards more ectomorphic body build was revealed in many studies, which was demonstrated by smaller values of chest circumference, other circumferences, body diameters, etc. (Godina 2001, Yampolskaya 2000).

To explain these changes, socio-economic conditions were mostly referred to, as positive secular changes in the tempos of growth and final adult body dimensions of populations were generally connected with improvements in environmental, dietary and living conditions, while the opposite - negative secular changes - were the result of a deterioration in the quality of life (Bogin 1999, Eveleth and Tanner 1990, Tobias 1985).

Another important source of intra group variability in growth and development is social stratification (Mascie-Taylor 1989). Due to political reason this subject was not very popular in Russia before "perestroika". However, in one of the early studies we correlated the values of some measurements and age at menarche in a sample of Moscow children investigated in the 1980's with parental education and occupation, and did not find any significant connection (Godina and Zadorojnaya 1990). Political and social changes of the 1990's increased the process of social stratification in Russia. It is a well-known fact that Moscow is a city with the large number of billionaires, while the general quality of life is declining (see e.g., commentary in the newspaper "Moskovsky komsomolets" on the review in "Forbes" magazine, May 20th 2004).

The aim of the current study is to assess the influence of family status and characteristics on parameters of children's growth and development, and to consider if there were any changes in those connections between 1980's and 1990's.

METHODS

Data used in the study were collected between 1996 and 2000. 1422 boys and 1530 girls from 7 to 17 were investigated cross-sectionally in several Moscow schools located in different city areas. The children were 98% Russians, mostly (60%) born in Moscow. They were measured during or immediately after school-hours; an age group consisted of children whose age falls within the interval $\pm$ 6 months of the whole year (e.g., 7-year olds: from 6.5 to 7.49, etc.).

A number of anthropometric measurements were taken according to standard techniques (Bounak 1941, Weiner and Lourie 1981). Subjects were measured bare-feet, wearing only underwear. Weight was taken using a battery-operated digital weighing scale (precision 100g).

Height and leg length (sum of the heights of iliospinale and symphysion points divided by two) were measured using an anthropometer (1mm precision). Body diameters (biiliac, bicristal, chest width and chest depth) were measured using a spreading caliper (1 mm precision). Chest, waist, arm, thigh and calf circumferences were measured using a plastic measuring tape (5 mm precision). Chest circumference was measured at the union of the 3rd and

4th sternebrae. Subcutaneous skinfolds (subscapular, biceps, triceps, abdomen, thigh and calf) were measured using a skinfold caliper, "Best modification" (1 mm precision).

The body mass index (BMI) of the studied subjects was calculated as weight (kg) divided by height $(m)^2$. Chest index was calculated as chest depth divided by chest width, %. Body proportions were calculated as corpus length divided by leg length, %.

Besides anthropometric measurements, stages of secondary sexual characteristics were assessed, age at menarche was evaluated by status quo method.

Questionnaire contained information on parental profession and education, family income, number of children in the family, birth order, birth weight and length of the investigated individual. For the present study only the information on father's and mother's education and the number of children per family was used. Three categories of parental education - separately for fathers and mothers - were chosen: 1 – school education (at least 8 classes); 2 – specialized secondary education; 3 – higher education. As for 'number of children per family' three groups were also formed: 1 – one child, 2 – 2 children, 3 – more than 2 children.

Obtained data were standardized with standard scores being calculated. Later it allowed summarizing the results according to different categories. Multiple regression analysis was used. Significance of differences was assessed by Scheffe test (one-way ANOVA).

Research was carried out in compliance with the Helsinki Declaration, approved by Bioethical Committee of Biological Faculty of Lomonosov Moscow State University.

RESULTS AND DISCUSSION

The number of children in each category isolated according to parental education was about 30% of the whole sample both for boys and for girls, while for the factor "number of children in the family", the last category "three children and more" comprised about 11-12% of the whole sample, not surprisingly for such a big city as Moscow.

The results of the correlation analysis show that the relation between the educational level of both parents is high: $r = 0.72$ in the girls' families, and $r = 0.68$ in the boys' families. As to the relationship between the educational level of parents and the number of children in the family the expected negative correlation was confirmed. The higher the father's and mother's education was, the smaller the number of children in the families. Correlation coefficients are rather small, but higher than in some other studies (Kromeyer et al. 1997, Mladenova 2003). For the category "father's education" the correlation is -0.20 for the boys, -0.26 for the girls. For mother's education and the number of children per family, it is -0.21 for both girls and boys.

The results of multiple regression analysis show that the contribution of social factors is more significant for boys than for girls: the values of determination coefficient (RI) are respectively 16% and 6-7% ($p < 0.0001$). The results can be explained by bigger "ecosensitivity" of boys comparing to girls, though in the Jena Study (Kromeyer et al. 1997) opposite results concerning gender differences have been received.

Figures 1, 2 and 3 show standardized residuals for the given factor after "removing" other factors included in multiple regression models. As can be seen, there are significant differences in morphological characteristics of the two sexes under the influence of different socioeconomic factors.

In boys there is an increase in the values of the majority of characteristics parallel to increase in father's education (Figure 1). For the girls the results are different. The biggest values of many measurements, particularly those relevant to body weight and fat component, were shown in group 1. Chest circumference and chest index (chest shape) were also bigger in the daughters of less educated fathers. There were no statistically significant differences for stature in girls of those three groups.

For mothers' education, the direction of differences in boys is practically the same as for father's education. Girls seem not responsive at all to the influence of this independent variable (no significant differences for any of the traits). (Figure 2). According to number of children per family, the direction of differences is the same: in both sexes body dimensions are decreasing parallel with increasing number of children. However, in girls the variations are more pronounced (Figure 3). It is also worth noting that the differences are significant mostly between the groups 1 and 2 for "father's education" category both for girls and for boys ($p < 0.05$), though the direction of the changes is different.

According to some literary data, the differences in morphological characteristics between boys and girls from the families of different educational and professional levels follow the same trend and are more pronounced in boys (Billewicz et al. 1983, Chinnock et al. 1984). In the case of Moscow children the situation is different. At the same time, average scores for educational level of parents in boys' and girls' families are practically the same: for boys, average score for “father’s education” is 2.12, “mother’s education” 2.05, “number of children…” 1.84; for girls, 1.93, 1.94 and 1.72 respectively.

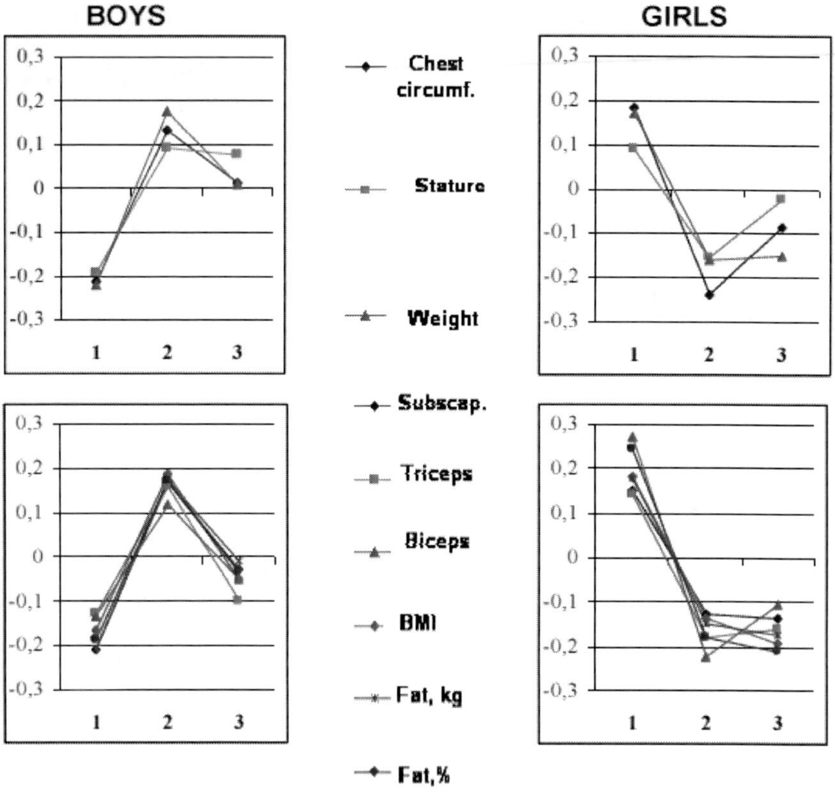

Figure 1. The influence of father’s education on some morphological characteristics in Moscow children (1 – school; 2 – secondary special; 3 – higher ed).

Thus, we can state distinct differences in the direction and degree of gender reactions towards social factors in Moscow children and adolescents of 1990’s. It can also be concluded that for children of both sexes higher education does not represent a factor of unambiguous increase in the level of somatic development as was shown in many other studies. The explanation of these results should be connected with social and economic changes in Russian society after “perestroika”. These results may be typical for other countries with “economies under transition”.

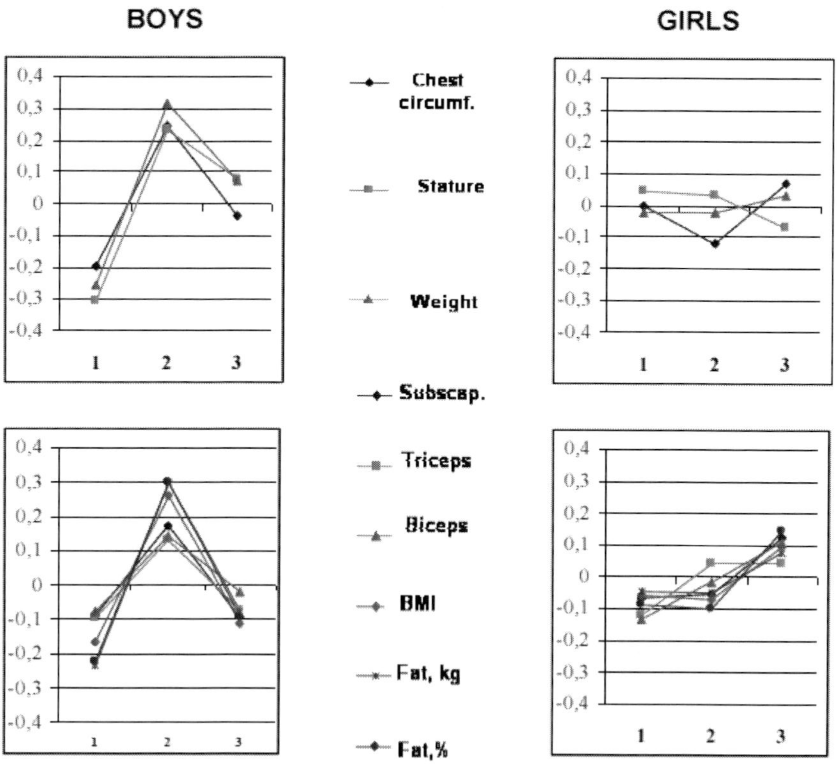

Figure 2. The influence of mother's education on some morphological characteristics in Moscow children (1 – school; 2 – secondary special; 3 –higher ed).

Variations in menarcheal age as connected with the level of father's education coincide with those for somatic traits. The lowest age at menarche at the girls from group 1 is 12.55. According to mother's education the results are different: the lowest age at menarche is at girls from the families with more educated mothers: in group 2 and 3, 12.21 and 12.29 respectively; and in group 1 it is 12.71. Thus, again we can see pronounced differences in the groups according to father's and mother's education. It may be speculated that the lowest age at menarche in a group with less educated fathers might be connected with a family stress (Hulanicka 1999, Hulanicka et al. 2001) but this hypothesis needs more profound research. With increase in the number of children per family menarcheal age is also increasing: 12.29, 13.09 and 13.44 years.

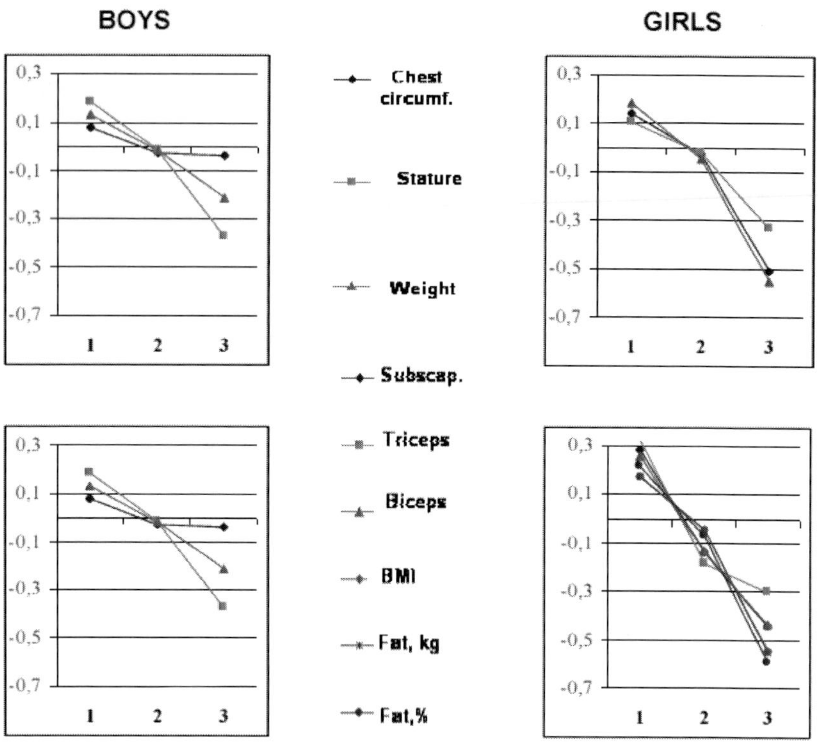

Figure 3. The influence of number of children per family on some morphological characteristics in Moscow children.

One of the aims of the present study was to reveal the influence of a school type on somatic development of children. A school, which the child attends, can be considered as indirect indication of a family status. In the last decade of the 20th century, along with ordinary municipal schools, a lot of new, "elite" schools (lyceums, gymnasia, boarding schools, etc.) emerged in and around Moscow. Some of these new schools are closed private institutions with high fees for children's education.

In Figures 4 and 5 average standard scores for a number of traits in children attending different schools are presented. Boys from municipal schools show much lower values in most measurements, while girls show lower body and leg length but higher body weight, trunk length, chest circumference and fat layer thickness, particularly on trunk (centripetal). Thus, the variability according to school type is parallel to that one analyzed according to the family status.

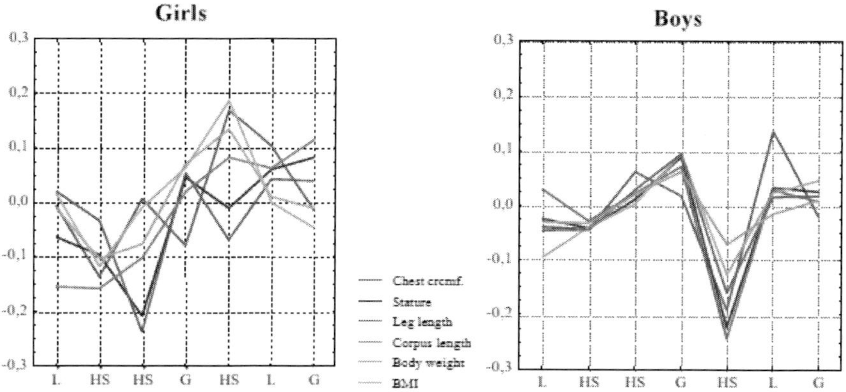

Figure 4. Variations of morphological characteristics in Moscow children from different schools. (L - lyceum, HS - high school, G – gymnasium).

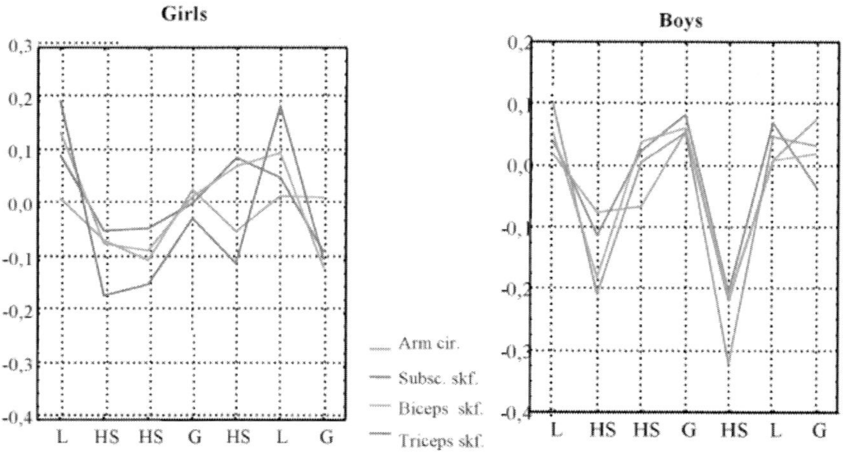

Figure 5. Variations of morphological characteristics in Moscow children from different schools (L – lyceum, HS – high school, G – gymnasium).

In one of the schools children from different classes were studied (Figure 6). In so called "lyceum" classes 180 boys and girls were examined, and in "ordinary" classes - 135. Again in many morphological characteristics the boys from "lyceum" classes surpass their counterparts from "ordinary" classes. The differences are significant for measurements in length - stature, leg length ($p< 0.05$), for body circumferences and skinfolds ($p < 0.05$).

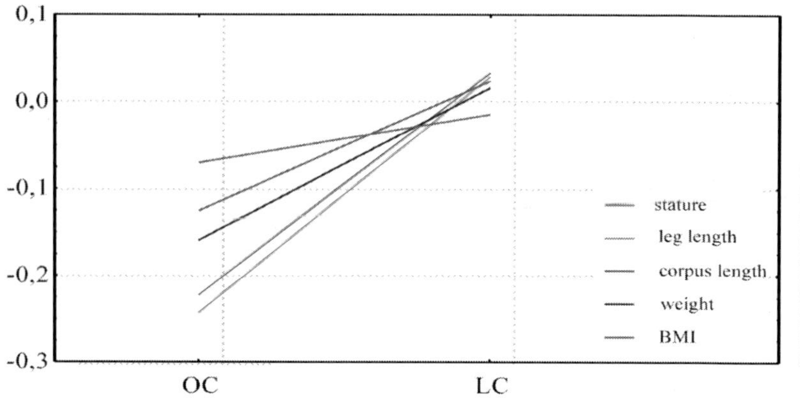

Figure 6. Variations of morphological characteristics in boys from different classes(OC ordinary classes, LC lyceum classes).

In girls the differences in stature are small and non-significant. Girls from "ordinary" classes have shorter legs, higher weight and BMI, larger values of biiliac diameter and chest depth, of chest circumference and subscapular skinfold. At the same time, the values of skinfolds on upper extremity are higher in girls from "lyceum" classes. The differences are significant for chest index and skinfold thickness ($p < 0.05$).

Certain differences are revealed between children of different classes in tempos of sexual maturation. As can be seen from Table 1, mean age of development of secondary sexual characteristics is in most cases higher for school children from ordinary classes. For menarcheal age the difference is two months in favour of "lyceum" girls. Similar results for girls attending the same school but different educational classes were reported by Grenier and Poniewiera (1983).

It is also interesting to evaluate secular changes in intragroup variations of growth and development parameters as influenced by socioeconomic factors. Table 2, based on the data by Plyaskina (1998) and Yampolskaya (2000), shows that in the 1990's the percentage of children with weight deficit was larger than in the 1970's, this tendency being particularly pronounced among children of "new" schools. These data fully confirm previously discussed results that in the last decade of the 20th century Moscow schoolchildren became more ectomorphic and asthenic (Godina 2001).

Table 1. Mean age of development of secondary sexual characteristics in boys and girls of different classes

Characteristics	Ordinary classes	Liceum classes
	GIRLS	
Ma	10 y. 0 m.	8 y. 11 m.
Ax	11 y. 8 m.	11 y. 6 m.
P	11 y. 6 m.	11 y. 8 m.
Me	13 y. 1 m.	12 y. 11 m.
	BOYS	
Ax	13 y. 9 m.	12 y. 11 m.
P	12 y. 8 m.	12 y. 3 m.

Table 2. Assessment of physical development (%) in Moscow schoolchildren in the 1970's and 1990's (Plyaskina, 1998; Yampolskaya, 2000)

Type of school/ year of investigation	N	Average	Excess of body weight	Deficit of body weight	Short stature
English school/ 1970's	611	81.8	11.6	4.9	1.7
Municipal school/ 1970's	709	79.0	10.5	8.6	1.9
Extended-day school/ 1970's	1150	79.5	5.2	13.3	2.0
Boarding school for special needs children/ 1980's	257	77.3	6.6	11.0	5.1
"New" schools/ 1990's	1080	73.8	8.2	16.7	1.3
Municipal schools/ 1990's	1003	75.0	10.0	12.0	3.0

The evidence of ongoing changes in social stratification can be seen in Figure 7. In the 1980's age at menarche did not vary due to parental education, in the 1990's as indicated above differences between groups are distinct.

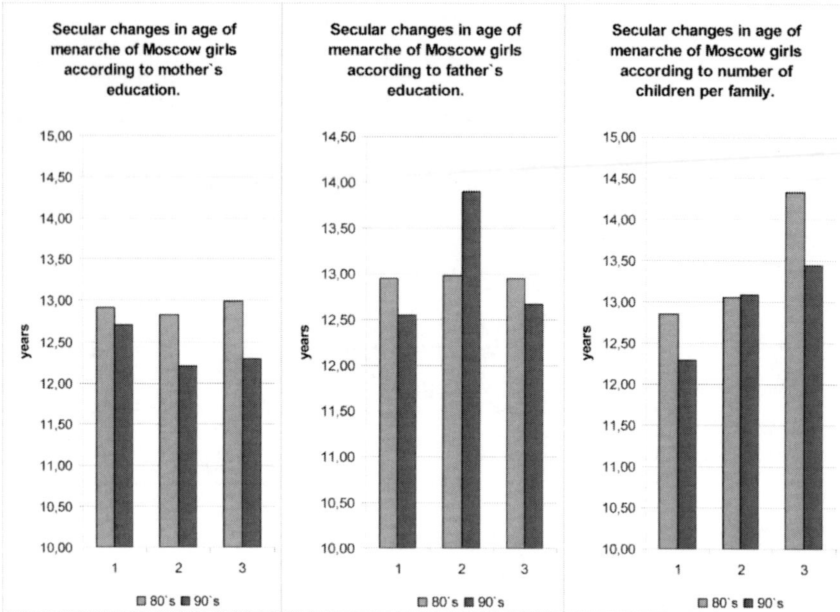

Figure 7. Changes in menarcheal age of Moscow girls

CONCLUSION

The results of the study show significant differences in somatic development of Moscow children due to their family status. For children of both sexes there are clear-cut variations in direction and degree of changes due to social factors. Parental higher education does not necessarily represent a factor of increase in the level of somatic development as was shown in many other studies. The explanation of these results should be connected with social and economic changes in Russian society after "perestroika". Secular changes repeat the tendencies revealed between different social groups. The impact of social stratification is becoming more evident in Russia and should be a matter of future research.

ACKNOWLEDGMENTS

The authors are grateful to Russian Foundation of Fundamental Research (RFFI) for their financial support and to all of the participants of this study.

REFERENCES

Billewicz, W. Z., Thomson, A. M., and H. A. Fellowes. 1983. "A longitudinal study of growth in Newcastle upon tyne adolescents." *Annals of Human Biology* 10:125-133.

Bogin, Barry. 1999. *Patterns of human growth.* 2nd Ed. Cambridge: Cambridge University Press, 1999.

Bounak, V. V. 1941. *Anthropometry.* Moscow: Uchpedgiz, (In Russian)

Chinnock, A., Reegan, T., Fox, P, et al. 1982. "Associations between growth patterns? Social factors, morbidity and developmental delay in a longitudinal survey of preschoolchildren." *Human Growth and Development*: 3rd International Congress on Auxology, Brussels, Aug. 26-30, 1982. N.Y., London, pp 699-703.

Eveleth, Phyllis B., and James M. Tanner. 1990. *Worldwide Variation in Human Growth.* 2nd Ed. Cambridge: Cambridge University Press.

Godina, Elena Z. 1998. "Secular changes in Russia and the former Soviet Union. In: Secular Growth Changes in Europe." Eds. E. Bodzsar and C. Susanne. Budapest: Eotvos Univ. Press, pp 351-367.

Godina, Elena Z. 2001. *Dynamics of growth changes in time and space.* D.Sc. thesis, Moscow, 2001.

Godina, Elena Z., and L. V. Zadorozhnaya. 1990. "The influence of some environmental factors on somatic development of children and adolescents (according to investigation of Moscow schoolgirls)." *Voprosy Antropologii* 84:18-30. (In Russian)

Grenier, E., and M. Poniewiera. 1983. "Contribution á l'etude de l'influence des conditions mesologiques sur la menarche". *Cahiers anthropologie et biometrie humaine* 1:25-83.

Hulanicka, B. 1999. Acceleration of menarcheal age of girls from dysfunctional families. *Journal of Reproductive and Infant Psychology* 17:119-132.

Hulanicka, Barbara, Gronkiewicz, Lidia, and Jan Koniarek. 2001. "Effect of familial distress on growth and maturation of girls: a longitudinal study." *American Journal of Human Biology* 13:771-776.

Kromeyer, K., Hauspie, Ronald Charles., Susanne, C. 1997. "Socioeceonomic factors and growth during childhood and early adolescence in Jena children." *Annals of Human Biology* 24:343-353.

Mascie-Taylor, C. G. Nicholas. 1989. "The biology of social class." In: Mascie-Taylor, C. G. N. (Ed). *Biosocial Aspects of Social Class*. Oxford: Oxford University Press.

Mladenova, S. N. 2003. *Anthropological characteristics of growth and development of children in Smolyan region*. Ph.D. thesis, Plovdiv. (In Bulgarian)

Plyaskina, I. V. Characteristics of the health status in schoolchildren of new types of educational institutions. Ph.D. thesis, Moscow. (In Russian)

Sukharev, A.G., and S. A. Mikhailova. 2000. The role of ecological and social factors in children's health status: the case of Altai Republic. Moscow: Ministry of Health, Russian Federation. (In Russian)

Tobias, Phillip V. 1985. "The negative secular trend." *Journal of Human Evolution* 14:347-356.

Weiner, Joseph D., and John A. Lourie. 1981. *Practical Human Biology*. London: Academic Press.

Yampolskaya, Y. A. Physical development of Moscow schoolchildren in the last decades. D. Sc. thesis, Moscow, 2000. (In Russian)

In: Biological Implications of Human Mobility ISBN: 978-1-63485-644-7
Editors: Slawomir Koziel et al.

Chapter 6

RURAL TO URBAN MIGRATION AND ITS IMPACT ON WOMEN'S HEALTH STATUS IN POLAND

Maria Kaczmarek [*]
Department of Human Biological Development, Institute of Anthropology, Faculty of Biology, The Adam Mickiewicz University, Poznań, Poland

ABSTRACT

Regional disparities in women's health and well-being in Poland have been a growing public health concern since the transformation to a market economy in the 1990s. This study aimed to examine the impact of internal migration on women's health after adjusting for demographic, (age, marital status, parity), reproductive (menstrual status), socioeconomic (education, occupation, income) and lifestyle behaviour (obesity, cigarette smoking, physical activity) covariates. The analyses of 10,777 study participants aged 20 to 80 based on women's self-rated health, number of chronic conditions and selected noncommunicable diseases revealed that women who migrated from rural villages to urban destinations were more likely to report better health, less likely to suffer from chronic diseases, and generally to be healthier than women who stayed in the village. At the same time however, large cities with their

[*] Corresponding Author address Maria Kaczmarek, Department of Human Biological Development, Institute of Anthropology, Faculty of Biology, The Adam Mickiewicz University, Poznań, Umultowska 89, 61-614 Poznań, Poland. E-mail: makac@amu.edu.pl.

various stressful stimuli appeared to be a serious risk factor of developing cardiovascular disease and other noncommunicable diseases, especially for younger women. The study results suggested that geographical mobility might be mediated through various combinations of demographic, socioeconomic and lifestyle behavioural factors. Additionally, the study confirmed that chronological age along with women's menopausal status is the most significant risk factor of health in the ageing women.

Comprehensive and in depth-knowledge of all potential consequences of rural-to-urban migration to women's health is of primary importance for policymakers and public health practitioners as well as for women themselves. This knowledge may be useful in developing effective public health policy interventions to reduce health inequalities and improve women's health.

Keywords: self-assessed health status, multiple chronic conditions, chronic diseases, urbanization, education, income adequacy

INTRODUCTION

According to the United Nations World Urbanization Prospects, in 2008 the world population reached a major demographic milestone sometimes referred to as the "Urban Millennium" or the "tipping point". For the first time in history the proportion of the population living in urban areas was equal the rural population (UN 2008:Figure I.1.). Since then the world has becoming urban in its majority. Today, over half of the total global population (54%) is living in urban areas and is projected to reach 66% by 2050 (UN 2015). Although the rapid growth of urban population occurs globally there is substantial variability in the levels of urbanization across regions (Figure 1).

At present, the most urbanized regions include Northern America with 81.5% people of its population living in urban areas, followed by Latin America and the Caribbean (79.5%), Europe (73.4%), and Oceania (70.8%). Africa and Asia, though mostly rural nowadays, are urbanizing faster than the other regions of the world - the proportion urban is increasing by 1.1% and 1.5% per annum, respectively as compared to regions that already have relatively high levels of urbanization – they are urbanizing at less than 0.4% per annum. These two continents are projected to become urban by 2030 with the proportion 53.5% and 54.5% of their respective populations (UN 2014).

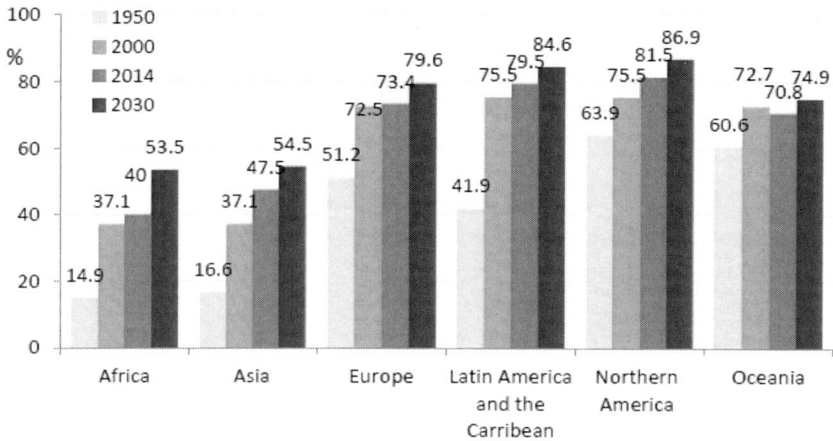

Figure 1. Urban population as a percent of total population by major areas, 1950, 2000, 2014 and projection for 2030. Source: UN 2014.

Global urban explosion is a recent phenomenon. In 1900 only 15% of the world's population lived in urban areas (UN 2008). This picture has been changed during the twentieth century as described above.

The Concept of Urbanization

Urbanization is an outcome of social development with primary dual meaning (Gist and Fava 1974, p.483). In one aspect, it is demographic, referring to the shift in the composition of the population measured by the urban population share, and the process by which that share is growing. In another aspect, it refers to the process whereby people are influenced by economic, social and environmental factors and the manner and pathways how people adjust and adapt to life in the biosocial urban environment (Lasker and Mascie-Taylor 1988, Blanquart et al. 2012, UN 2013).

Urbanization, however defined, is much more than a simple matter of urban dwellers or goods accounting (Knox 1994). The concept of urbanization refers to a dynamic (temporal), multifaceted phenomenon that may be approached from various viewpoints: geographic, demographic, anthropological, health status, economic, political, legal, cultural, religious, and social – to mention but a few (Pumain 2006). Urbanization occurs through series of interlinked transformations driven by three interacting processes: the net movement of people from rural to urban areas, reclassification of rural settlements into towns and cities and the natural population increase i.e., when birth rate exceeds death rate (Gold 1982, Weeks 2008, Nauman et al. 2016). The net migration from rural to urban areas is a key determinant of changes in population distribution and a major driving force in the process of urbanization. The rural-to-urban migration is always accompanied by territorial and occupational reorganization in that it shifts land-use and shifts from agricultural to non-agricultural occupations, as well as changes in demographic structure and in many aspects of life in a given society (McGranahan and Satterthwaite 2014, White 2015). The process of urbanisation via population growth can occur directly when natural increase in rural population exceeds a given threshold and the population becomes reclassified as urban, and indirectly via higher fertility rates among rural-to-urban migrants (Nauman et al. 2016).

Urbanization and Health

Internal migration is usually accompanied by social upward mobility as by achieving higher level of education, changing jobs, livelihood and lifestyle behaviour, and often marrying (Gold 1982, UN 2011). The movement up the socioeconomic ladder results in an array of biological changes in the growth and development, fertility, morbidity and mortality (Bogin 1988). Uppwardly mobile people are likely to be taller, healthier and generally better biologically and physically fit than their less well-to-do peers from rural areas (Bielicki and Charzewski 1983, Bielicki and Waliszko 1992, Lipowicz 1999, Welon et al. 1999, Bielicki et al. 2003). Rural-urban pattern of health differences involves also a higher tendency towards premature death among adult rural dwellers (Kołodziej et al. 2008). However, the health impact of urbanisation is double-sided. On the one hand, there are the benefits of the accessibility of resources, health services, sanitation *and* secure nutrition in urban areas, while on the other hand the increased density of urban population and diversity of cities are associated with higher risk of psychosocial stressors. Exposure to these psychosocial hazards can affect people's health, both physical and mental, through stress-mediated pathways (Pickett and Pearl 2001, Moorea et al. 2003). There is ample evidence of literature that recent urbanization and accompanying enormous changes in social, economic, cultural and environmental domains are the leading drivers of the increase in prevalence of noncommunicable diseases (Moorea et al. 2003, Leon 2008, Allender et al.

2011). Furthermore, profound changes in nutrition (particularly increases in dietary saturated fat intake), more sedentary behaviour and physical inactivity are likely to increase risk of obesity and chronic illness such as type II diabetes, hypertension and cardiovascular diseases (Abraído-Lanza et al. 2005, Ebrahim et al. 2010, Masterson Creber et al. 2010, Unwin et al. 2010, Bowen et al. 2011, Chan et al. 2012, Hernandez et al. 2012, Carrillo-Larco et al. 2016).

The exact mechanisms through which the process of urbanization may affect human health have not been well understood, there is however suggestive evidence on the intermediate socioeconomic, psychosocial and behavioural pathways linking migration status with health outcomes (Lu 2010). A better understanding of pathways in which environmental stimuli act and how they can affect health status becomes an increasingly important issue in public health policy (Gushulak et al. 2010).

Internal Migration in Poland

Poland, as a typically rural country with late industrialization, has witnessed different patterns and paces of internal migration since the Second World War (Fallenbuchl 1977). In the wake of the Second World War, many villagers moved to cities. The largest rural-to-urban migration took place during the period 1950-1970. Other waves involved 1980s and 1990s (Fallenbuchl 1977). The share of people living in rural areas has subsequently changed over time. In the 1950s, 60% of the population lived in rural areas. The proportion urban was 50% in 1965 and continued to fall dramatically until 2002 thus representing 38% of the total population (Golinowska et al. 2006). The share of rural dwellers amounted to 38.5% in 2004 (total population of Poland 38 173 835 people and 14 703 749 living in villages) and 39.7% in 2014 (total population of Poland 38 478 602 people and 15 262 250 living in villages) (GUS 2015).

The internal migration resulted also in the age-structural differences between urban and rural areas. People living in rural areas were likely to be much older than their urban counterparts. The high total fertility rate (TFR) of the urban population was another factor affecting demographic disproportions between these two populations. The high TFR in urban areas began to fall in the 1950s and stabilized at the reproduction level. This main trend *in the family structure and dynamics has changed considerably since the beginning of the transition from* socialist *centralized* system mainly based on central planning *to a market economy* in the 1990s. The *birth rate has* largely *been declining* since then due to *a rapid decline in fertility rates and changes in fertility patterns.*

The total fertility rate (TFR) has been dropping steadily since 1990. In 2015 it amounted 1.33. The mean age at first child birth for women increased from 23.7 years in 2000 to 27.2 years in 2013 (GUS 2015, EUROSTAT 2016).

Why Women's Health Matters?

According to GUS data, in 2014 women in Poland represented a little over half (51.6%) of the total population (GUS 2015). C*ensus* figures indicate that Poland has made a remarkable progress in increasing life expectancy since 1950 but yet it is one of the lowest life expectancies for both women and men among European Union countries (OECD 2015). Life **expectancy at birth** (LE_0) **for women born in Poland in 2013 was** 81 years, and had increased of almost twenty years since 1950 (OECD 2015). There is gender gap in LE_0 suggesting that women can expect to live over 8 years longer than men whose LE_0 is estimated 73 years. In 2013 LE at age 65 (LE_{65}) was estimated to reach 18.9 years among women and 14.5 years among men indicating that gender differences amounted to 3-4 years (OECD 2015).

The gender gap in health status is quite prominent if using objective measures such as LE_0 and LE_{65} with women living longer than men. However, these additional years of women's life seem to be associated with their worse health as shown by subjective self-reported health indicators such as disability adjusted life years (DALYs) and health adjusted life expectancy (HALE). In Poland, in 2013, HALE was 70.03 years (95%CI: 66.77– 72.88 years) for women and 63.86 years (95%CI: 61.35– 66.20 years) for men (GBD 2013 DALYs and HALE Collaborators 2015). These figures indicate that women may outlive men but they spend a comparatively lesser part of their lives without diseases and disabilities (86.4%). By contrast, men can expect to live 87.7% of their lifespan free of disabilities. The four most common causes of DALYs include (in rank order): ischemic heart diseases (IHD), stroke, low back and neck pain and malignant neoplasms (GBD 2013 DALYs and HALE Collaborators 2015).

The leading causes of death among Polish women include IHD, cerebrovascular diseases, colorectal, breast and lung cancers (Krzyżak et al. 2015). According to predicted cancer mortality for the EU, breast and colorectal cancers have promising (falling) trends, but predicted 9% lung cancer rates rise is becoming the highest rate cancer, possibly overtaking breast cancer rates (Malvezzi et al. 2015).

Inequality in education, occupation and income exacerbates large inequalities in life expectancy, morbidity and mortality. Women with the highest level of education tend to live six years longer on average than those with the lowest level (OECD 2015).

In conclusion, it may be said that women's health is a subject of major interest because women constitute half the population in Poland, live longer than men and encounter more health challenges; they perform multiple roles including the roles of wife and employee, caring for children and elderly relatives, providing primary health care and making decisions about health care for themselves and their family. To function in these roles they must be healthy themselves.

Study Objectives

The main objective of this study was to ascertain the impact of urbanization upon the health of Polish women after controlling for demographic, reproductive (menstrual status), socioeconomic and cultural covariates.

Our working hypothesis was that urbanization would be a predictive factor of women's health regardless their age.

The following research questions were addressed: i) Do rural-to-urban migrants differ in health status from their counterparts who stayed in the village? ii) Is rural-to-urban migration associated with increased hazards for noncommunicable diseases? iii) To what extend chronological age and women's menstrual status are factors differentiating their health outcomes?

This paper consists of two sections: the first section presents decription of women's health status on the basis of self-report estimates across age groups and levels of menstrual status (premenopausal versus postmenopausal) and the association of urbanization, socio-economic and lifestyle behaviour factors with health estimates. The second section presents prevalence of multiple chronic conditions and seven most frequently reported noncommunicable diseases in relation to urbanization, socio-economic and lifestyle behaviour factors after controlling for age and menstrual status.

METHODS

Participants, Study Design and Instruments

The data used in this study emanate from three cross-sectional surveys: the national survey of women's health and well-being in midlife (WOMID) on 7,243 women aged between 35 and 65 years and conducted in 2000–2004. This survey has already been described in detail elsewhere (Kaczmarek 2007). Two community-based surveys on 2,223 women aged between 20 and 35 years and on 1,311 women aged 60 and 80 years conducted in 2005 and 2006, respectively have not been published yet. The total sample consisted of 10,777 women ranged in age from 20 to 80, represented all social strata and resided in rural and urban areas in all regions of the country.

The study was designed in a cross-sectional, personal *interview* survey by which the researcher administered the questionnaire in a *face-to-face* manner. A Polish version of menopause-specific questionnaire (MSQ) was administered to participants of the WOMID study (Kaczmarek 2000) and self-reported socio-demographic, lifestyle behaviour, reproductive history and medical history data were collected. Additionally, a Polish version of the Medical Outcomes Study Short Form 36-item Questionnaire (SF-36) instrument was used to assess perceived health status (Ware and Sherbourne 1992). Self-assessment of health status was based on the first question in the SF-36 instrument "How do you assess your health status?" and on a five-point Likert-type scale, where 1) represented very poor, 2) poor, 3) fair, 4) good and 5) very good and/or excellent self-rated health status.

Basic anthropometric measurements such as body height and weight were taken according to standard procedures (Knussmann, 1988/1992). Then, body mass index (BMI) was calculated by taking a subject's weight (kg) and dividing it by her height squared.

Variables

All analyses performed in this study were adjusted to chronological age and age-related physiological changes in menstrual status resulting in being either premenopausal or postmenopausal. The menstrual status served as a proxy for age.

Chronological age was calculated in decimal values by subtracting the date of examination from the date of birth and was classified into four age groups: 20 to 35 years, 35 to 50 years, 50 to 65 years and 65 to 80 years.

The natural menopause was referred by a woman retrospectively as to the last menstrual period (LMP) in her life resulting from cessation of ovarian follicular activity. The postmenopausal status was defined as amenorrhea for 12 or more consecutive months dating from the last menstrual period, for which there were no other obvious pathological or physiological cause for menses cessation. The premenopausal status included women who had regular menstruation during the last 12 months (WHO 1981, NHAMS 2010). Women with surgical menopause were excluded from analysis.

Health outcomes included self-rated health status (SRH), morbidity described in terms of multiple chronic conditions (MCCs) and seven most frequently reported diseases such as cardiovascular disease (CVD), hypertensive disease (HTND), respiratory diseases, neurotic disorders, urogenital diseases, type 2 diabetes and malignant neoplasms. The focus was on how these conditions vary according to urbanization factor and other social, economic, cultural and demographic covariates. There is no standard definition for MCCs. Many studies define MCCs as 2 or more conditions, and some, mostly for clinical purposes, define MCCs as 3 or more conditions. In this population-based study we adopted the definition of 2 or more conditions (Diederichs et al. 2011).

Place of residence was determined on the basis of the settlement typology in Poland (Rogacki 2007) and stratified to four levels of living areas: rural areas (village with population of less than 1000 inhabitants, mainly engaged in farm work and this work is a source of income), small town with less than 20,000 inhabitants, medium-sized city with 20,000 to 100,000 inhabitants and large city with more than 100,000 inhabitants.

The internal migration was defined in terms of four categories of the movement from the village to town/city with the refrence group of women who were born in the village and lived their lives there versus those who were born in the village and moved to i) small town, ii) medium-sized city and iii) large city. Rural-to-urban migrants had to live in the urban destination areas for at least 10 years.

The demographic factor related to women's health status included chronological age and current marital status with never married/cohabiting as a reference category (married/cohabiting, widowed, divorced/separated).

The social roles of women related to their health status were characterized by the parity meant as the number of children ever born to a woman with one child category being referenced (2, 3, 4+ children) and occupational status. Three types of employment status included women who were unemployed as a reference category versus women economically inactive – a category that includes women who are pension, retired, those raising a family at home and women who voluntary remain out of the active workforce and those who were employed.

The socio-economic status was constructed on the basis of educational attainment, employment status (of double meaning for both social roles and socioeconomic status) and financial position. Educational attainment was classified into three levels according to the education system in Poland with primary/vocational with less than 12 years of schooling as a reference category versus secondary level with 12 years of schooling and tertiary/high level with more than 12 years of education. Financial position of women was described in terms of income adequacy as an ordinal measure of not enough money (a reference category), just enough and more than enough to cover expenses each month.

Lifestyle behaviour was assessed in terms of weight status, tobacco use and physical activity. Women with healthy weight (BMI between 18.5 to 24.9 kg/m^2) served as the reference category for comparisons with their overweight (25-29.9 kg/m^2) and obese ($\geq$30 kg/m^2) counterparts. Never smoked women (reference category) were compared to former and current smokers. Physical activity during leisure time was self-reported by a woman in terms of number of hours spent on physical exercising and then categorized as physically active ($\geq$1h per week) or non-active (a reference category).

Data Analysis

The dependent variable was women's perceived health status and selected chronic diseases. Potential predictors of health outcomes were independent variables including urbanization (categorical variable), marital status (categorical variable), parity (categorical variable), elements of socioeconomic status such as educational attainment, occupational and financial status (categorical variables), and lifestyle behaviour such as weight status (either a continuous or categorical variable), tobacco use and physical activity level (both categorical variables). All analyses were adjusted to age and menstrual status.

Descriptive statistics were used to describe the basic features of the study sample including means and 95% confidence interval, standard deviations and percentage frequency histograms.

Multiple logistic regression analysis (MLR) was used to assess the association between perceived health status and its potential predictors. The odds ratio and its confidence interval were used as a measure of association. A final explanatory model with a subset and odds ratios (OR) of factors associated with health status was obtained using a stepwise procedure with backward elimination and rejection criterion of the p-value greater than 0.05.

Cox proportional hazards regression model was used in analysing a binary outcome in a cross-sectional study according to suggestions from Lee and Chia (1993). The hazard ratio and its confidence interval were estimated under an assumption that the hazard ratio was constant over time. This hazard ratio (HR) or "relative risk" is the estimated hazards for a case with disease to those without disease and indicates the change in the risk of the event (in our study the occurrence of disease) when there is a change of one unit in the independent variable. If the relative risk for a given variable category is bigger than 1 it means that the occurrence of this category value is associated with a higher risk of the occurrence of a disease compared with the reference category. In our model, women without the disease under analysis were treated as censored, meaning that it was not known when they would get ill, but at least until the time of the survey they were well. Those reporting a disease were treated as completed. The hazard rate function was estimated for selected noncommunicable diseases such as cardiovascular disease, hypertensive diseases, respiratory diseases and neurotic disorders to express how the hazard rate depends upon a set of demographic, socioeconomic and lifestyle behaviour covariates.

Statistical analyses were performed using the STATISTICA 12 software (StatSoft Polska Sp. z o.o.). To determine whether the difference was statistically significant, the t-test for continuous variables and chi-square test for categorigal variables were used. All significance tests comprised two-way determinations. A value of $p < 0.05$ was considered statistically significant.

Table 1. Demographic and socioeconomic characteristics of study participants stratified by place of residence

Variable	Village n = 2274	Small town n = 4321	Mid-sized city n = 1811	Large city n = 2371	*p*-value[a]
Age (years)	50.48±9.48	50.37.±8.58	49.92±8.65	51.01±9.27	0.014
Marital status					<0.001
Never Married/Cohabiting	89 (3.9)	255 (5.9)	112 (6.2)	152 (6.4)	
Married/Cohabiting	1815 (79.8)	3323 (76.9)	1371 (75.7)	1804 (76.1)	
Widowed	300 (13.2)	510 (11.8)	186 (10.3)	251 (10.6)	
Divorced/Separated	70 (3.1)	233 (5.4)	142 (7.8)	161 (6.9)	
Parity					<0.001
Nulliparous	84 (3.7)	320 (7.4)	147 (8.1)	154 (6.5)	
Parous 1 child	273 (12.0)	851 (19.7)	427 (23.6)	628 (26.4)	
2 children	889 (39.10	2169 (50.2)	933 (51.5)	1211 (51.1)	
3 children	641 (28.2)	929 (21.5)	344 (18.8)	393 (16.6)	
4 and more	471 (20.7	372 (8.6)	110 (6.1)	139 (5.9)	
Education level, years of schooling					<0.001
Primary/Vocational <12 years	1248 (54.9)	1322 (30.6)	442 (24.4)	461 (19.4)	
Secondary 12 years	751 (33.0)	2191 (50.7)	907 (50.1)	1038 (43.8)	
Tertiary/Higher >12 years	275 (12.1)	808 (18.7)	462 (25.5)	872 (36.8)	
Employment status					<0.001
Unemployed	400 (17.6)	501 (11.6)	176 (9.7)	275 (11.6)	
Economically inactive[b]	662 (29.1)	1149 (26.6)	398 (22.0)	396 (16.7)	
Employed	1212 (53.3)	2671 (61.8)	1237 (68.3)	1700 (71.7)	
Income adequacy[c]					<0.001
Not enough	753 (33.1)	1176 (27.2)	464 (25.6)	519 (21.9)	
Enough	1282 (56.4)	2536 (58.7)	1034 (57.1)	1463 (61.7)	
More than enough	239 (10.5)	609 (14.1)	313 (17.3)	389 (16.4)	
Weight status (BMI)[d]					<0.001
Healthy weight	1014 (44.6)	2087 (48.3)	993 (54.8)	1245 (52.5)	
Overweight	784 (34.5)	1573 (36.4)	595 (32.8)	849 (35.8)	
Obesity	475 (20.9)	661 (15.3)	223 (12.4)	277 (11.7)	
Smoking status					<0.001
Never smoked	1196 (52.6)	1979 (45.8)	844 (46.6)	1110 (46.8)	
Former smoker	348 (15.3)	825 (19.1)	319 (17.6)	372 (15.7)	
Current smoker	730 (32.1)	1517 (35.1)	648 (35.8)	889 (37.5)	
Physical activity					<0.001
Non active	1890 (83.1)	3448 (79.8)	1380 (76.2)	1605 (67.7)	
Active, ≥1h/week	384 (16.9)	868 (20.1)	431 (23.8)	766 (32.3)	

Mean±SD for continuous variables; n(%) for categorical variables.

[a] *p*-values for place of residence differences derived from chi-square Pearson test for categorical variables and Student *t*-test for continuous variables.

[b] A category that includes people who are pension, retired, those raising a family at home and people who voluntary remain out of the active workforce.

[c] An ordinal measure of more than enough, just enough, or not enough money to cover expenses each month.

[d] Healthy weight: BMI<25 kg/m^2; Overweight: BMI = 25-29.9 kg/m^2; Obesity: BMI > 30 kg/m^2.

RESULTS

Characteristics of Study Participants

Demographic, socioeconomic and lifestyle behaviour characteristics of study participants stratified by their place of residence are shown in Table 1.

Women currently living in villages consisted of 21.1% of the total sample and those living in small towns accounted for 40.1%. Seventeen percent of women lived in medium-sized cities and 22% in large cities. Substantial differences in socio-economic and lifestyle behaviour characteristics were found along rural-urban gradient with a tendency of the advantage to locate in urban areas. Women from urban areas, unlike their rural counterparts, had more often a university degree (36.8%, 25.5%, 18.7%, versus 12.1% for large, medium-sized city, small town and village, respectively), were more often employed (71.7%, 68.3%, 61.8% versus 53.3%, respectively) and wealthier (16.4%, 17.3%, 14,1% versus 10.5% of women reporting that have more than enough money to cover expenses each month in large, medium-sized city, small town and village, respectively). Families in cities and towns were smaller in size than those in the country having rarely 4 or more children (5.9%, 6.1%, and 8.6% versus 20.7% in large, medium-sized city, small town and village, respectively). Healthy weigth and physical activity were more commonly reported by women residing in large cities accounting for 52.5%, 54.8%, 48.3% versus 44.6% of healthy weigth women and 32.3%, 23,8%, 20.1% versus 16.9% of physically active women in respective category of residency). At the same time however, women from urban settings were more frequently current smokers than their counterparts from rural areas with 37.3% of current smokers in large, 35.8% in medium-sized cities and 35.1% in small towns versus 32.1% of their counterparts residing in villages.

With regard to women's menstrual status, half of them (50.4%) were premenopausal and the other half (49.6%) were postmenopausal. Median age at menopause was 51.25; $Q_1 = 49$; $Q_2 = 54$ and did not reveal significant difference between rural and urban dwellers (Kaczmarek 2007, 2007a). Mean age of premenopausal women was 43.6 years SD = 4.56 years and those postmenopausal was 57.41 years SD = 6.65 years.

The urbanization process in the study sample is depicted in Figure 2. Of all the women born in village, 38.2% continued to live their lives in rural settings while others moved to the city. One fourth of women (26.2%) born in village moved to small towns. Those who moved to large cities accounted for 11.6%. The median length of living in urban settings for rural-to-urban migrants was 13 years. Women aged 20-25 years who had spent less than 10 years in towns or cities were excluded from analysis on urbanization factor.

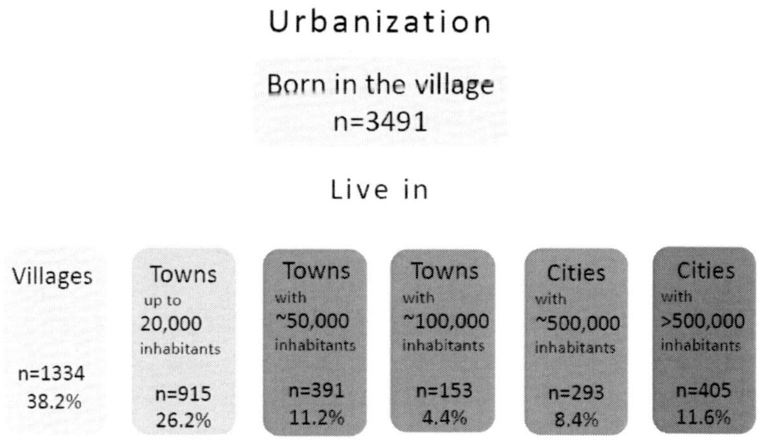

Figure 2. Distribution of various categories of destination place for women born in village.

Self-Rated Health Status

Self-rated health status (SRH) showed a strong relationship with age approving that age was the most significant demographic factor for health ($\chi^2 = 256.1$, df = 12, $p < 0.001$) (Figure 3). SRH got steadily worse with age. The share of women assessing their health as very bad tends to increase with continuous lines from the youngest to the oldest groups of women. In the group of women aged 65-80 it accounted for 11% as compared to zero percent in the youngest group of women aged 20-35. Self-rated health was shown to be less than good for 77.4% of women aged 65-80 and for 34.8% of women aged 20-35.

The distribution of responses to SRH by current place of residence revealed consistent rural-to-urban gradient ($\chi^2 = 301.4$, df = 12, $p < 0.001$) (Figure 4). Women living in villages were twice more often than their counterparts residing in large cities to assess their health as very bad (3.9% versus 1.9% of women from villages and large cities, respectively). Regardless of the place of residence, over 60% of women tend to assess their health as less than good. There was almost constant proportion of women assessing their health status as very good varying from 3.2% of large city residents and women living in countryside to 2.9% -2.8% of small town and medium-sized city dwellers. These findings highlight pronounced differences in health assessment across rural and urban settings and the critical role the place of residence plays in women's health.

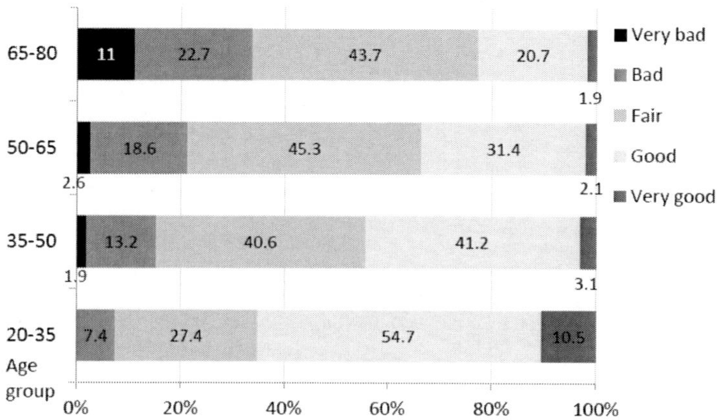

Figure 3. Percentage distribution of women's self-rated health scores according to age group.

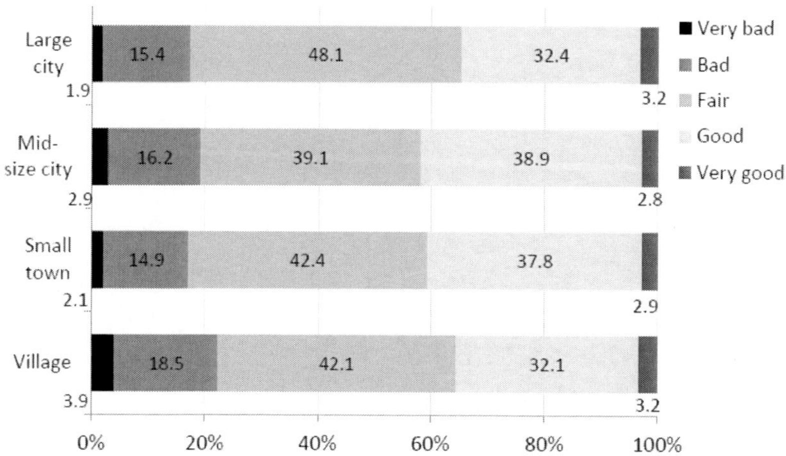

Figure 4. Percentage distribution of women's self-rated health scores according to place of residence.

Table 2. Multiple logistic regression most parsimonious set of factors affecting the likelihood of self-rated less than good health in premenopausal and postmenopausal women: odds ratios and 95% confidence interval

Stepwise MLR using backward selection Self-rated less than good health status			
Premenopausal women[a]		Postmenopausal women[b]	
Variable	OR (95% CI)	Variable	OR (95% CI)
Urbanization		Urbanization	
Ref. Village-Village	1	Ref. Village-Village	1
Village-Small town	0.82*(0.61;0.97)	Village-Small town	0.87*(0.81;0.95)
Village-Mid-sized city	0.74**(0.62;0.89)	Village-Mid-size city	0.70**(0.62;0.84)
Village-Large city	1.12*(0.97;1.44)	Village-Large city	1.09*(0.90;1.59)
Education level		Employment status	
Ref. Primary/Vocational	1	Ref. Unemployed	1
Secondary	0.72 (0.56;0.93)	Economically inactive	0.71 (0.58;0.87)
Tertiary/higher	0.53 (0.32;0.87)	Employed	0.51 (0.33;0.71)
p-value for trend	0.011	*p*-value for trend	0.0008
Employment status		Income adequacy	
Ref. Unemployed	1	Ref. Not enough	1
Economically inactive	0.77 (0.59;0.99)	Enough	0.76 (0.55;0.79)
Employed	0.59 (0.36;0.98)	More than enough	0.62 (0.51;0.83)
p-value for trend	0.031	*p*-value for trend	<0.0001
Income adequacy			
Ref. Not enough	1		
Enough	0.81 (0.68;0.91)		
More than enough	0.66 (0.48;0.91)		
p-value for trend	0.011		
Weight status (BMI kg/m^2)			
Ref. Healthy weight	1		
Overweight	1.32 (1.04;1.66)		
Obesity	1.73 (1.09;2.74)		
p-value for trend	0.018		

Adjusted for urbanization category, marital status, parity, education level, employment status, income adequacy, weight status, smoking status, and physical activity.

[a] Chi-square = 64.02 df = 5; $p < 0.0001$.

[b] Chi-square = 65.61; df = 3; $p < 0.0001$.

Statistically significant at: *$p < 0.05$; **$p < 0.01$.

A set of factors affecting the likelihood of SRH in premenopausal (chronologically younger) and postmenopausal (chronologically older) women, derived from data adjusted to all study covariates is shown in Table 2. It included urbanization, employment status, educational attainment and income adequacy in both premeneopausal and postmenopausal women and additionally weight status in premenopausal women. An interesting pattern of association between SRH (with reference category of less than good health) and urbanization was found. Women who had moved from village to small town and medium-size city were likely to rate their health better than their non-migrant counterparts who stayed in the village (OR = 0.82; 95%CI:0.61;0.97 and OR = 0.74; 95%CI:0.62;0.89 in premenopausal women and OR = 0.87;95%CI:0.81;0.95 and OR = 0.70;95%CI:0.62;0.84 in postmenopausal women, for respective category of urbanization factor). By contrast, women who had moved from village to large city were more likely than their counterparts who did not move to assess their health worse (odds ratios were OR = 1.12; 95%CI:0.97;1.44 and OR = 1.09; 95%CI:0.90;1.59).

Women with a tertiary/higher level of education, in paid employment and very good income (having more than enough money to cover expenses each month) were almost twice less likely than those unemployed with income inedaquacy and low level of education to assess their health as less than good. By contrast, premenopausal obese women were almost twice as much likely as their healthy weight counterparts to asees their health as less than good (OR = 1.73; 95%CI:1.09;2.74).

Morbidity

The number of women who have not just a single but multiple co-occurring chronic conditions (MCCs) increased significantly with age (χ^2 = 224.7, df = 3, $p < 0.001$). This substantial increase of MCCs with advancing age is shown in Figure 5. The prevalence of MCCs showed almost threefold increase across age varying from 20.5% in women aged 20-35, through 26.9% in women aged 35-50, 44.7% in women aged 50-65, and 70.9% in women aged 65-80.

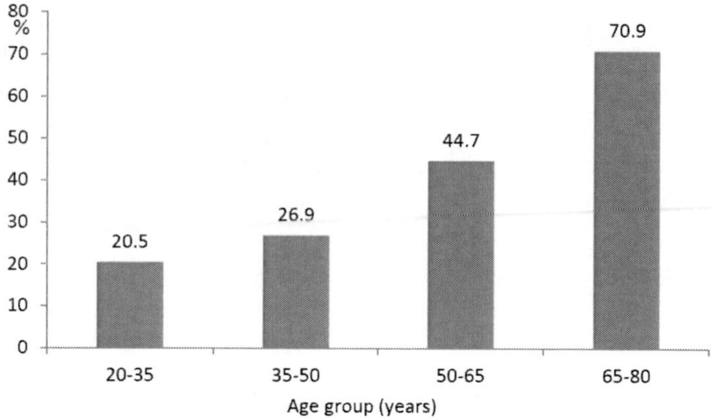

Figure 5. Prevalence (%) of reported multiple chronic conditions among women by age group.

The prevalence of MCCs also varied according to place of residence (χ^2 = 14.8, df = 3, $p < 0.01$) in continuous line from village (47.6%), through small town (42.8%), medium-size city (36.8%) to large city (34.8%) as shown in Figure 6. The greatest difference was found between women residing in villages versus those from large cities (47.6% versus 34.8%).

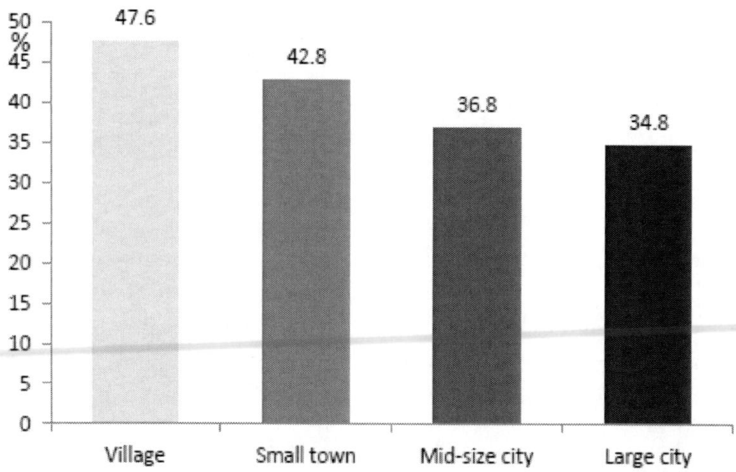

Figure 6. Prevalence (%) of reported multiple chronic conditions among women by place of residence.

Table 3. Cox proportional hazards regression most parsimonious models of risk factors associated with multiple chronic health conditions in premenopausal and postmenopausal women: hazard ratios and 95% confidence interval

Stepwise Cox proportional hazards regression using backward selection Multiple (≥2) chronic health conditions			
Premenopausal women[a]		Postmenopausal women[b]	
Variable	HR (95% CI)	Variable	HR (95% CI)
Urbanization		Urbanization	
Ref. Village-Village	1	Ref. Village-Village	1
Village-Small town	1.68* (0.49;5.70)	Village-Small town	1.14* (1.01;1.71)
Village –Mid-sized city	0.46** (0.16;1.24)	Village –Mid-sized city	0.82* (0.76;1.55)
Village –Large city	0.86 (0.27;1.24)	Village –Large city	0.76*(0.69;1.09)
Employment status		Marital status	
Ref. Unemployed	1	Ref. Never Married/Coh.	1
Economically inactive	0.49 (0.31;0.876	Married/Cohabiting	1.33* (1.17;2.15)
Employed	0.32**(0.23;0.46)	Widowed	2.14**(1.93;3.05)
		Divorced/Separated	1.23* (0.98;1.95)
Income adequacy			
Ref. Not enough	1		
Enough	0.60**(0.47;0.78)		
More than enough	0.43**(0.21;0.66)		
Weight status (BMI kg/m^2)			
Ref. Healthy weight	1		
Overweight	1.49*(1.23;1.96)		
Obesity	1.69*(1.13;2.32)		
Smoking status			
Ref. Never smoked	1		
Former smoker	1.28 (0.86;1.85)		
Current smoker	1.88**(1.38;2.56)		
Physical activity			
Non active	1		
Active, ≥1h/week	0.64**(0.46;0.88)		

Adjusted for urbanization category, marital status, parity, education level, employment status, income adequacy, weight status, smoking status, and physical activity.
[a] 2Log likelihood (constant) = 2883.78; AIC = 2915.78; Wald = 117.34.
[b] 2Log likelihood (constant) = 366.79; AIC = 398.79; Wald = 15.14.
Statistically significant at: *$p < 0.05$; **$p < 0.01$.

Adjusted hazards for having a risk score of two or more versus zero or one chronic disease revealed several risk factors including urbanization in both pre- and postmenopausal women and employment, income adequacy, and lifestyle behaviour factors such as health status, smoking habit and physical activity exclusively in pre-menopausal women whereas marital status exclusively in postmenopausal women (Table 3).

Migrants from village to small town were at almost twofold higher risk of developing MCCs than their counterparts who stayed in the village (HR = 1.68, 95%CI 0.49;5.70 in premenopausal women and HR = 1.14, 95%CI 1.01;1.71 in postmenopausal women, $p < 0.05$). By contrast, women who had moved from village to medium-sized and large cities proved to be at lower risk of developing MCCs than their counterparts who stayed in the village (HR = 0.46, 95%CI 0.16;1.24 and HR = 0.82, 95%CI 0.76;1.55 for pre- and postmenopausal women moving to medium-sized cities and HR = 0.76, 95%CI 0.69;1.09 for premenopausal women moving to large cities).

Widowed (HR = 2.14, 95%CI 1.93;3.05), obese (HR = 1.69, 95%CI 1.13;2.32) and currently cigarette smoking women (HR = 1.88, 95%CI 1.38;2.56) were more likely to be at higher risk of developing MCCs than their never married, healthy weight and never smoked counterparts.

The analysis revealed constantly protective effects of socioeconomic advantage against MCCs hazards including employment (HR = 0.32, 95%CI 0.23;0.46), good financial position (HR = 0.43, 95%CI 0.21;0.66) and physical activity (HR = 0.64, 95%CI 0.46;0.88).

Among younger women (aged 20-35 years) with 2 or more MCCs, the most prevalent dyad was neurotic disorders/duodenal and gastric ulcers. This dyad was also one of the most prevalent among those aged 35-50 followed by chronic bronchitis and pneumonia/urogenital diseases. Among older women those aged 50-65 and 65-80 who had 2 or more chronic diseases, the most prevalent dyad was ischemic heart disease/hypertension and genitourinary disease/respiratory disease, respectively.

The increased prevalence of selected diseases with age is depicted in Figure 7. The prevalence of CVD, HTND and respiratory diseases (Figure 7A) exhibited a tenfold rise between the age groups 20-35 and 65-80 varying from 5.4% to 46.7%, 6.5% to 49.6% and 1.1% to 11.7% in case of CVD, HTND and respiratory diseases, respectively. The estimation of age-related prevalence of neurotic disorders indicated subtle upward trend varying from 16.3% in women aged 20-35 to 20% in those aged 65-80.

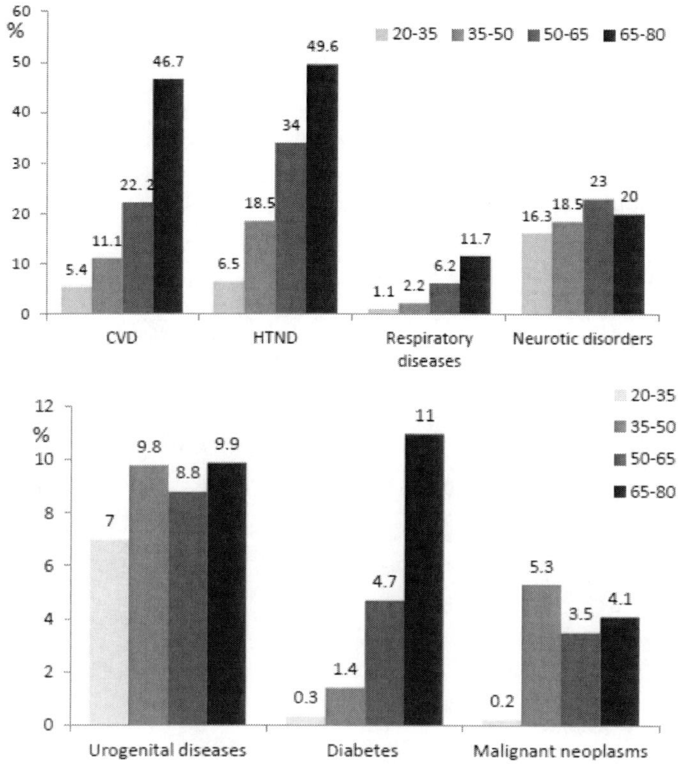

(A)

(B)

Figure 7. Prevalence (%) of main reported noncommunicable diseases among women by age group: A) cardiovascular disease (CVD), hypertensive diseases (HTND), respiratory diseases, neurotic disorders B) urogenital diseases, diabetes, malignant neoplasms.

A similar subtle upward trend was found for urogenitial diseases. The prevalence of urogenital diseases varied from 7% in women aged 20-35 to 9.9% in those aged 65-80. By contrast, prevalence of type 2 diabetes and malignant neoplasms substantially increased with age varying from 0.3%, 0.2% in women aged 20-35 to 11% and 4.1% in women aged 65-80 in relation to type 2 diabetes and malignant neoplasms, respectively (Figure 7B).

Cardiovascular Disease

Adjusted hazards of developing CVD revealed a set of hazard factors such as urbanization and employment status in both pre- and postmenopausal women and marital status, weight status and cigarette smoking status exclusively in postmenopausal women (Table 4).

Table 4. Cox proportional hazards regression most parsimonious models of cardiovascular disease risk factors in premenopausal and postmenopausal women: hazard ratios and 95% confidence interval

Stepwise Cox proportional hazards regression using backward selection Cardiovascular disease			
Premenopausal women[a]		Postmenopausal women[b]	
Variable	HR (95% CI)	Variable	HR (95% CI)
Urbanization		Urbanization	
Ref. Village-Village	1	Ref. Village-Village	1
Village-Small town	0.44* (0.21;0.89)	Village-Small town	1.33 (0.83;2.14)
Village –Mid-sized city	0.35** (0.18;0.69)	Village –Mid-sized city	1.12 (0.75;1.69)
Village –Large city	0.79 (0.37;1.67)	Village –Large city	0.66**(0.26;0.49)
Employment status		Employment status	
Ref. Unemployed	1	Ref. Unemployed	1
Economically inactive	1.23* (1.07;1.69)	Economically inactive	0.58 (0.39;0.87)
Employed	1.47* (0.96;1.94)	Employed	0.36**(1.51;2.29)
Weight status (BMI kg/m^2)		Marital status	
Ref. Healthy weight	1	Ref. Never Married/Coh.	1
Overweight	1.40*(1.22;3.02)	Married/Cohabiting	1.39 (0.42;4.58)
Obesity	1.79*(1.46;3.09)	Widowed	2.44**(1.07;5.56)
		Divorced/Separated	1.55 (0.67;3.59)
Smoking status			
Ref. Never smoked	1	Weight status (BMI kg/m^2)	
Former smoker	1.17 (0.95;1.83)	Ref. Healthy weight	1
Current smoker	1.66**(1.29;2.05)	Overweight	1.20*(1.12;2.17)
		Obesity	1.66*(1.35;1.99)
Physical activity			
Non active	1	Smoking status	
Active, ≥1h/week	0.56**(0.35;0.79)	Ref. Never smoked	1
		Former smoker	1.06*(0.75;1.51)
		Current smoker	1.78**(1.31;2.17)

Adjusted for urbanization category, marital status, parity, education level, employment status, income adequacy, weight status, smoking status, and physical activity.

[a] 2Log likelihood (constant) = 2689.61; AIC = 2699.61; Wald = 103.47.

[b] 2Log likelihood (constant) = 2867.72; AIC = 2893.72; Wald = 102.1.

Statistically significant at: *$p < 0.05$; **$p < 0.01$.

The risk of developing CVD was twofold lower for premenopausal (younger) rural to urban migrants and postmenopausal women who had moved from village to large city as compared to those who stayed in the village with respective hazard ratio estimates HR = 0.44; 95%CI 0.21;0.89, HR = 0.35; 95%CI 0.18;0.69 and HR = 0.66; 95%CI 0.26;0.49.

Attention should be drawn to the fact that the employment status both increased and reduced the risk of developing CVD. Younger and employed women were in 1.48 (95%CI 0.96;1.94) higher risk of developing CVD than their unemployed counterparts. In older women however, employees were at threefold lower risk of developing CVD (HR = 0.36; 95%CI 1.51;2.29) than their unemployed counterparts.

The risk of developing CVD was higher for premenopausal obese (HR = 1.79; 95%CI 1.46;3.09), cigarette smoking (HR = 1.66; 95%CI 1.29;2.05) women with protective effect of physical activity (HR = 0.56; 95%CI 0.35;0.79. Postmenopausal widowed, obese and cigarette smoking women as compared to their never married, healthy weight and never smoked counterparts were likely to be at higher risk of developing CVD with respective hazard ratio estimates HR = 2.44; 95%CI 1.07;5.56, HR = 1.66; 95%CI 1.35;1.99 and HR = 1.78; 95%CI 1.31;2.17.

Hypertensive Diseases

Adjusted risk factors of developing HTNDs revealed that younger, obese and cigarette smoking migrants from village to large city were likely to be at substantially higher risk of this disease than their healthy weight, never smoked counterparts who stayed in the village (Table 5). In postmenopausal women, the following risk factors were identified: urbanization, employment and income adequacy, marital status, weight status and smoking habit. Widowed (HR = 2.61; 95%CI 1.07;3.21), obese (HR = 1.43; 95%CI 1.11;1.83), and cigarette smoking (HR = 1.68; 95%CI 1.24;2.16), migrants from village to large city (HR = 1.27; 95%CI 0.96;1.54) were at substantially higher risk of developing HTND than their never married, healthy weight and never smoked counterparts who did not move out from village. Employment status and income adequacy both reduced the risk of HTND. Employed, wealthy women were at four times and two times lower risk of developing HTND than their jobless, poor counterparts with respective hazard ratio estimates HR = 0.28; 95%CI 0.21;0.37 and HR = 0.61; 95%CI 0.42;1.57.

Respiratory Diseases

Full adjusted hazards of respiratory diseases (RDs) included demographic, socioeconomic and lifestyle behaviour factors (Table 6).

Table 5. Cox proportional hazards regression most parsimonious models of hypertensive disease risk factors in premenopausal and postmenopausal women: hazard ratios and 95% confidence interval

Stepwise Cox proportional hazards regression using backward selection Hypertensive diseases			
Premenopausal women[a]		Postmenopausal women[b]	
Variable	HR (95% CI)	Variable	HR (95% CI)
Urbanization		Urbanization	
Ref. Village-Village	1	Ref. Village-Village	1
Village-Small town	0.94 (0.69;1.34)	Village-Small town	0.97* (0.83;1.14)
Village –Mid-sized city	0.88 (0.68;1.19)	Village –Mid-sized city	0.89* (0.75;1.19)
Village –Large city	1.29* (0.97;1.47)	Village –Large city	1.27*(0.96;1.54)
Employment status		Marital status	
Ref. Unemployed	1	Ref. Never Married/Coh.	1
Economically inactive	0.83 (0.69;1.05)	Married/Cohabiting	1.53*(0.71;2.38)
Employed	0.28**(0.21;0.37)	Widowed	2.61**(1.07;3.21)
		Divorced/Separated	2.24*(1.05;4.08)
Income adequacy			
Ref. Not enough	1	Weight status (BMI kg/m^2)	
Enough	0.84* (0.61;1.25)	Ref. Healthy weight	1
More than enough	0.61**(0.42;1.57)	Overweight	2.01**(1.48;2.67)
		Obesity	1.43*(1.11;1.83)
Weight status (BMI kg/m^2)			
Ref. Healthy weight	1	Smoking status	
Overweight	1.57**(1.02;1.99)	Ref. Never smoked	1
Obesity	2.33**(2.02;5.18)	Former smoker	1.06 (0.85;2.51)
		Current smoker	1.68**(1.24;2.16)
Smoking status			
Ref. Never smoked	1		
Former smoker	1.25*(0.98;1.49)		
Current smoker	1.47*(1.05;1.74)		

Adjusted for urbanization category, marital status, parity, education level, employment status, income adequacy, weight status, smoking status, and physical activity.

[a] 2Log likelihood (constant) = 3150.37; AIC = 3154.37; Wald = 124.29.

[b] 2Log likelihood (constant) = 3573.48; AIC = 3589.48; Wald = 129.52.

Statistically significant at: *$p < 0.05$; **$p < 0.01$.

Table 6. Cox proportional hazards regression most parsimonious models of respiratory disease risk factors in premenopausal and postmenopausal women: hazard ratios and 95% confidence interval

Stepwise Cox proportional hazards regression using backward selection Respiratory diseases			
Premenopausal women[a]		Postmenopausal women[b]	
Variable	HR (95% CI)	Variable	HR (95% CI)
Urbanization		Urbanization	
Ref. Village-Village	1	Ref. Village-Village	1
Village-Small town	0.96 (0.89;1.64)	Village-Small town	0.99 (0.81;2.14)
Village –Mid-sized city	0.98 (0.78;1.16)	Village –Mid-sized city	1.02 (0.95;1.89)
Village –Large city	0.89* (0.67;1.07)	Village –Large city	0.97 (0.86;3.56)
Marital status		Marital status	
Ref. Never Married/Coh.	1	Ref. Never Married/Coh.	1
Married/Cohabiting	1.35*(0.97;1.64)	Married/Cohabiting	1.42*(1.22;2.08
Widowed	0.93 (0.69;2.54)	Widowed	1.82**(1.57;2.21)
Divorced/Separated	0.92 (0.82;3.21)	Divorced/Separated	1.32*(1.15;2.08)
Employment status		Smoking status	
Ref. Unemployed	1	Ref. Never smoked	1
Economically inactive	1.03 (0.89;1.65)	Former smoker	1.28* (1.14;1.67)
Employed	0.58**(0.41;0.87)	Current smoker	1.30*(1.22;2.11)
Income adequacy		Weight status (BMI kg/m^2)	
Ref. Not enough	1	Ref. Healthy weight	1
Enough	0.84* (0.61;1.25)	Overweight	2.62**(1.12;1.85
More than enough	0.68**(0.54;0.86)	Obesity	2.04**(1.51;2.72)
Smoking status			
Ref. Never smoked	1		
Former smoker	1.42**(1.18;2.79)		
Current smoker	1.64**(1.25;2.94)		
Weight status (BMI kg/m^2)			
Ref. Healthy weight	1		
Overweight	1.23*(1.12;1.49)		
Obesity	1.83**(1.62;3.21)		

Adjusted for urbanization category, marital status, parity, education level, employment status, income adequacy, weight status, smoking status, and physical activity.

[a] 2Log likelihood (constant) = 5150.07; AIC = 5174.07; Wald = 64.29.

[b] 2Log likelihood (constant) = 4040.05; AIC = 4072.05; Wald = 51.99.

Statistically significant at: *$p < 0.05$; **$p < 0.01$.

In premenopausal women, those who were married/cohabiting, obese and currently smoking cigarettes were likely to be at significantly increased risk of RDs than never married/cohabiting, healthy weight and never smoked women (HR = 1,35; 95%CI 0.97;1.64, HR = 1.83; 95%CI 1.62;3.21 and HR = 1.64; 95%CI 1.25;2.94, respectively). The movement from village to large city (HR = 0.89; 95%CI 0.67;1.07) and in paid employment (HR = 0.58; 95%CI 0.41;0.87) were two factors that significantly reduced the risk of developing RDs as compared to unemployment and living in the village.

Postmenopausal, widowed, obese and currently cigarette smoking women regardless of their category of urbanization were likely to be at substantially higher risk of developing RDs than their never married/cohabiting, never smoked, healthy weight counterparts. The respective hazard ratio estimates were: HR = 1,35; 95%CI 0.97;1.64, HR = 1.83; 95%CI 1.62;3.21 and HR = 1.64; 95%CI 1.25;2.94.

Table 7. Cox proportional hazards regression most parsimonious models of neurotic disorders risk factors in premenopausal and postmenopausal women: Hazard ratios and 95% confidence interval

Stepwise Cox proportional hazards regression using backward selection Neurotic disorders			
Premenopausal women[a]		Postmenopausal women[b]	
Variable	HR (95% CI)	Variable	HR (95% CI)
Urbanization		Urbanization	
Ref. Village-Village	1	Ref. Village-Village	1
Village-Small town	1.04 (0.99;2.16	Village-Small town	1.04 (0.99;2.16
Village –Mid-sized city	1.01 (0.89;1.21)	Village –Mid-sized city	1.84* (0.69;1.01)
Village –Large city	0.99 (0.80;1.24)	Village –Large city	0.89* (0.76;0.99)
Marital status		Marital status	
Ref. Never Married/Coh.	1	Ref. Never Married/Coh.	1
Married/Cohabitating	1.43*(1.24;1.83	Married/Cohabiting	1.12 (1.03;1.98
Widowed	1.03 (0.84;1.54)	Widowed	1.63**(1.37;1.99)
Divorced/Separated	1.02 (0.92;2.21)	Divorced/Separated	1.42**(1.12;1.85)
Employment status			
Ref. Unemployed	1		
Economically inactive	0.89 (0.53;1.05)		
Employed	0.27**(0.15;0.49)		

Adjusted for urbanization category, marital status, parity, education level, employment status, income adequacy, weight status, smoking status, and physical activity. Statistically significant at: $*p < 0.05$; $**p < 0.01$. [a] 2Log likelihood (constant) = 1030.05; AIC = 1068.05; Wald = 62.58.
[b] 2Log likelihood (constant) = 949.79; AIC = 1981.79; Wald = 51.85.

Neurotic Disorders

Marriage/cohabiting status was a significant risk factor of neurotic disorders in premenopausal women. Married/cohabiting women as compared to never married/cohabiting counterparts were likely to be at significantly higher risk of neurotic disorders (HR = 1.43; 95%CI1.24;1.83). Employment played a protective role against neurotic disorders. The risk of developing neurotic disorders in premenopausal women was four times lower for employees than unemployed women (HR = 0.27; 95%CI 0.15;0.49).

In postmenopausal women a set of risk factors included urbanization and marital status indicating that currently employed migrants from village to large city were at significantly lower risk of developing neurotic disordes than their counterparts who stayed in the village (HR = 0.89; 95%CI 0.76;0.99). Widowed, divorced/separated women were at higher risk of neurotic disorders that their never married counterparts (HR = 1.63, 95%CI1.37;1.99 and HR = 1.42, 95%CI1.12;1.85 for widowed and divorced/separated, respectively).

DISCUSSION

In Poland, regional disparities in health status and well-being have been growing public health concern since the transformation to a market economy in the 1990s (Zatoński 1996, Golinowska et al. 2006, Boyle 2011).

The present contribution shows a comprehensive and in depth study on the impact of rural-to-urban migration on women's health status. The analyses based on women's self-rated health, number of chronic conditions and selected noncommunicable diseases demonstrated that women who migrated from rural villages to urban destinations were much less likely to report the chronic disease problems and more likely to report better general health than women who did not move from village. Migrating women were likely to have higher level of education, employment status and higher income than their non-migrating counterparts. *This is an evidence of* a *complex interactive relationship among geographic, demographic and socio-economic variables suggesting that* socio-economic and lifestyle behavior factors may translate women's health disparities associated with their movement from rural to urban areas. Large cities however with chronic crowding, noise and air pollution induce stressful stimuli which cause a wide variety of adverse health effects including CVD and other noncommunicable diseases. This creates an ambiguous picture of potential benefits of urbanization process to women's health.

The findings of this study support the thesis that both chronological age and menstrual status are substantial for women's health. It is well accepted that the timing of menopause may serve as a borderline for women's health. There is epidemiologic evidence that women who go through menopause before age 46 are twice as likely to suffer from ischemic heart disease, osteoporosis or have stroke whereas menopause after age 55 is associated with higher risk of breast, ovarian, and endometrial cancers (Dossus et al. 2010; Wellons et al. 2012; CGHFBC 2012). Postmenopuasal women are likely to worsen their well-being due to adverse clinical symptoms which usually accompany menopause, particularly those of vasomotoric nature (Dennerstein et al. 2000).

It is interesting to note that parity, though significantly different between rural and urban dwelling women did not affect women's health upon adjustment for other covariates in multivariate models. It suggests that the potential impact of parity on women's health may be blurred by other socioeconomic factors such as employment status and income adequacy.

Another interesting finding suggests that elements of socioeconomic status such as employment and income are more important for health of younger (premenopausal) women than their older counterparts (see Tables 4 through 7). Marital status becomes of primary importance for health of older (postmenopausal) than younger (premenopausal) women. These findings corroborate well those of Wróblewska (2002) who found similar pattern of relationship between SES and health status in Polish women.

Findings of this study confirm recent studies which have demonstrated geographical and socioeconomic inequalities in health across European countries. A comparative study of the magnitude of inequalities in self-assessed health among 22 European countries revealed that substantial variation of these inequalities were associated with socioeconomic status and lifestyle behaviours such as smoking, obesity, excessive alcohol consumption, and deficiencies in health care (Mackenbach et al. 2008). The study results have suggested that inequalities in access to good-quality health care have a role in generating inequalities in survival from chronic conditions.

The study of Helasoja and colleagues (2006) identified similar socioeconomic health patterns in both the Baltic countries (Estonia, Latvia, Lithuania) and other European countries, suggesting that the lower educated people had worse mental and physical health. This finding was confirmed by Rėklaitienė and colleagues (2008) in a sample of 263 rural women aged 24-77 and examined in 2006 in Lithuania. A cross-sectional survey undertaken in 2000 in Ukraine on a representative national sample of 1,600 adults revealed that as much as 43% of women rated their health as poor or very poor. This rating was worse than levels recorded in Russia and considerably worse than levels seen in Western Europe. Women living in villages were at 3.2-fold increased risk of poor self-rated health (OR = 3.24;95%CI1.30;8.07) than their counterparts living in cities (Bobak et al. 1998, Gilmore et al. 2002).

Rural-urban disparities in health and well-being were reported by Bińkowska (2005). She found in her study conducted in 2004 on a sample of 1,083 Polish women aged 45-54 that women living in rural areas had a significantly worse health than their urban counterparts. The same pattern of rural-urban disparities in health were found in a study of 2,143 peri- and postmenopausal women aged 45-65, dwellers of the Lublin province conducted by Żołnierczuk-Kieliszek and colleagues in 2007 (2012). The study conducted in 2009 by Zagozdon and colleagues (2011) on a sample of 1,560 Polish women aged 45-60 also revealed a large rural-urban gap in health-related quality of life in disfavour of rural dwelling women.

A significant association of residential area with women's health-related outcomes and psychological well-being on disfavour of rural dwelling women was reported by author in her previous works (Kaczmarek 2004, Kaczmarek and Skrzypczak 2012).

Wróblewska (2002) using data from 1996 Health Survey in Poland of over 20,000 women showed the impact of the new socioeconomic situation in the transition to a market economy on women's health and well-being. She found..."The association between women's health and their economic situation, marital status and other roles, as well as life style variables, such as obesity, smoking and physical activity.." (p. 723).

The HAPIEE cross-sectional study of random population samples in Novosibirsk (Russia), Kraków (Poland), Kaunas (Lithuania) and six Czech towns revealed pronounced social gradients in functional limitations consistent with previous findings elsewhere (Doryńska et al. 2012).

The morbidity pattern determined in this study is consistent with a new trend of growing number of people with multiple chronic diseases (Karvonen-Gutierrez 2015). Furthermore, it confirmed the leading role of CVD in women's health burden. Women tend to be diagnosed with CVD somewhat 7-10 years later in life than men and are more likely to die from MI than men (Towfighi et al. 2009). This sex-based difference in CVD pathophysiology has been primarily explained by estrogen loss during the menopausal transition. Estrogen is believed to have protective effects against coronary heart disease

(CHD) and other atherosclerotic diseases. Most likely mechanism of this protection is through improvement of serum HDL concentration and improvement of endothelial functions. It seems however, that the explanation is more complex than only that attributed to the loss of estrogen during the menopausal transition (Bhupathy et al. 2010). Socioeconomic status, social relationships, lifestyle behaviour such as smoking, drinking alcohol, lack of excercises, poor diet and obesity are modifiable factors as important to women's health as biological risk factors associated with menopausal transition and reproductive and physiological ageing. There are several hypotheses that may explain inequlities in health associated with migration. One of the most frequently offered explanations is the health selection hypothesis also known as *healthy migrant hypothesis*. It states that migrants represent a selectively healthy group that is different from their comaptriots who do not migrate and therefore stands out when they are compared with the general population at destination (Palloni and Morenoff 2001). When applied to our study it explains that young women of better health are migrating to urban settings leaving those of worse health in countryside, and they contribute to rural-urban health inequalities. Another possible explanation within the framework of ecological approach indicates that if the quality of host environment is adavatageous for newly comming people it supports them in their health (McLaren and Hawe 2005).

Apart from this general explanation, there are several causal pathways that have been hypothesized for understanding the mechanisms that translate urbanization process to health disparities at the community and individual levels. They include unequal availability and accessibility of resources and health services; infrastructure deprivation; attitudes towards health and the adoption of health maintenance behaviours; psychosocial mechanisms and a lack of social support (Mackenbach et al. 2008, Pickett and Pearl 2001). Although it is not established which of these pathways matter more for women's health, they each are likely to be intimately connected with one another or influence each other in reciprocal ways that accentuate on health effects. Factors at the geographic and social levels affect individual health behaviours such as the mode of spending leisure time, substance use, healthy dietary habits and individual psychological characteristics such as attitudes to health and self-esteem. All these factors in an aggregate manner significantly contribute to disparities in women's health.

This study has several weaknesses including the cross-sectional design which makes it difficult to assess the direction and causality. This design however is methodologically appropriate for evaluating the association between health outcomes and exposures (rural and urban areas, SES, lifestyle behaviours) (Mann 2003). Another weak point of this study may derive from errors in recall of the exposure and possible outcome may produce a reporting bias. The reliability of self-reported data has widely been discussed in the literature and was taken into account in this study (Fan et al. 2006).

Another bias may be produced by errors in recall of the exposure and possible outcome. It would have been useful to have longitudinal, prospective information. The reliability of self-reported data has widely been discussed in our previous study and involved in the premises of this study

The strengths of this study include a population-based design survey with the proportion of the population in rural and urban areas and social strata reflecting the nationwide proportions and the multivariate approach which allows us to integrate multiple factors and confounders hypothesized as health covariates.

Conclusion

The present study confirmed the association between internal migrations from rural to urban areas and women's health. It also revealed a set of factors such as demographic, *socioeconomic* and lifestyle behaviour factors mediating health disparities between women from different categories of urbanization.

A comprehensive and in depth-knowledge of all potential consequences of rural-to-urban migration to women's health is of primary importance for policymakers and public health practitioners as well as for women themselves. This knowledge may be useful in developing effective public health policy interventions to reduce health inequalities and improve women's health.

References

Abraído-Lanza, A.F., Chao, M.T., Flórez, K.R. 2005. "Do healthy behaviors decline with greater acculturation? Implications for the Latino mortality paradox." *Social Science and Medicine* 61:1243–55.

Allender, S., Wickramasinghe, K., Goldacre, M., Matthews, D., Katulanda, P. 2011. "Quantifying urbanization as a risk factor for noncommunicable disease." *Journal of Urban Health: Bulletin of the New York Academy of Medicine* 88(5):906-18.

Bhupathy, P., Haines, C.D., Leinwang, L.A. 2010. "Influence of sex hormones and phytoestrogens on heart disease in men and women." *Womens Health (Lond Engl)* 6(1):77-95.

Bielicki, T., Charzewski, J. 1983. "Body height and upward social mobility." *Annals of Human Biology* 10(5):403-408.

Bielicki, T., Waliszko, A. 1992. "Stature, upward social mobility and the nature of statural differences between social classes." *Annals of Human Biology* 19(6):589-93.

Bielicki, T., Szklarska A., Kozieł S., Welon, Z. 2003. *Socio-economical transition in the light of anthropological survey of 19-year-old males.* Monographs of the Institute of Anthropology PAN, No 23.

Bińkowska, M., 2005. "Badanie postaw i zachowań zdrowotnych oraz jakość życia reprezentatywnej grupy kobiet polskich w wieku od 45 do 54 lat." *Przegląd Menopauzalny* 4 supl.(1):1-72.

Blanquart, F., Gandon, S., Nuismer, S.L. 2012. "The effects of migration and drift on local adaptation to a heterogeneous environment." *Journal of Evolutionary Biology* 25: 1351–63

Bobak, M., Pikhart, H., Hetzman, C., Rose, R., Marmot, M. 1998. "Socioeconomic factors, perceived control and self-reported health in Russia. A cross-sectional survey." *Social Science and Medicine* 47(2): 269-79.

Bogin, B. 1988. "Rural-to-Urban Migration" In *Biological Aspects of Human Migration*, edited by C.G.N. Mascie-Taylor and G.W. Lasker. New York, New York, Cambridge University Press, 90-129. (Cambridge Studies in Biological Anthropology Volume 2.)

Bowen, L., Ebrahim, S., De Stavola B., Ness, A., Kinra, S., Bharathi, A.V. et al. 2011. "Dietary intake and rural-urban migration in India: a cross-sectional study." *PLoS One* 6:e14822.

Boyle, P. 2011. "Improving Health in Central and Eastern Europe." *Annals of Agricultural and Environmental Medicine* 18(2), 281-282.

Carrillo-Larco, R.M., Bernabé-Ortiz, A., Pillay, T.D., Gilman, R.H., Sanchez, J.H., Poterico, J.A., Quispe, R., Smeeth, L., Miranda, J.J. 2016. "Obesity risk in rural, urban and rural-to-urban migrants: prospective results of the PERU MIGRANT study." *International Journal of Obesity* 40:181-5.

Chan, F., Adamo, S., Coxson, P., Goldman, L., Gu, D., Zhao, D. et al.2012. "Projected impact of urbanization on cardiovascular disease in China." *International Journal of Public Health* 57:849–54.

Collaborative Group on Hormonal Factors in Breast Cancer. 2012. "Menarche, menopause, and breast cancer risk: individual participant meta-analysis, including 118 964 women with breast cancer from 117 epidemiological studies." *Lancet Oncology* 13:1141-51.

Dennerstein, L., Dudley, E.C., Hopper, J.L., Guthrie, J.R., Burger, H.G. 2000. "A prospective population-based study of menopausal symptoms." *Obstetrics and Gynecology* 96:351–8.

Diederichs, C., Berger, K., Bartels, D.B. 2011. "The measurement of multiple chronic diseases — a systematic review on existing multimorbidity indices." *Journals of Gerontology Series A Biological Sciences and Medical Sciences* 66(3):301–11.

Doryńska, A., Pająk, A., Kubinova, R., Malyutina, S., Tamosiunas, A., Pikhart, H., Peasey, A., Nikitin, Y., Marmot, M., Bobak, M. 2012. "Socioeconomic circumstances, health behaviors and functional limitations in older persons in four Central and Eastern European populations." *Age Ageing* 41:728–35.

Dossus, L., Allen, N., Kaaks, R. et al. 2010. "Reproductive risk factors and endometrial cancer: the European Prospective Investigation into Cancer and Nutrition." *International Journal of Cancer* 127:442–51.

Ebrahim, S., Kinra, S., Bowen, L., Andersen, E., Ben-Shlomo, Y., Lyngdoh, T. et al. "The effect of rural-to-urban migration on obesity and diabetes in India: a cross-sectional study." *PLoS Medicine* 7:e1000268.

Fallenbuchl, Z.M. 1977. "Internal Migration and Economic Development under Socialism: The Case of Poland" In *Internal Migration A Comparative Perspective* edited by A.A. Brown, E. Neuberger, 305-28. Academic Press, New York, N.Y., United States.

Fan, X., Miller, B. C., Park, K., Winward, B. W., Christensen, M., Grotevant, H. D. 2006. "An exploratory study about inaccuracy and invalidity in adolescent self-report surveys." *Field Methods* 18:223–44.

Gilmore, A.B., Mckee, M., Rose, R., 2002. "Determinants of and inequalities in self-perceived health in Ukraine." *Social Science and Medicine* 55(12):2177–88.

Gist, N.P., Fava, S.F. 1974. *Urban Society*. 6th ed. New York, NY: Thomas Y. Crowell.

Gold, H. 1982. *The Sociology of Urban Life*. Prentice Hall PTR.

Golinowska, S., Sowa A., Topór-Mądry, R. 2006. "Health and Morbidity in the Accession Countries Country Report – Poland." ENEPRI Research Report No. 29/November 2006. Accessed January 12, 2016 http://www.enepri.org.

Gushulak, B.D., Weekers J., MacPherson, D.W. 2010. "Migrants and emerging public health issues in a globalized world: threats, risks and challenges, an evidence-based framework." *Emerging Health Threats Journal* 2:e10. doi: 10.3134/ehtj.09.010 & 2010 BD.

Helasoja, V., Lahelma, E., Prättälä, R., Kasmel, A., Klumbiene, J., Pudule, I. 2006. "The sociodemographic patterning of health in Estonia, Latvia, Lithuania and Finland." *European Journal of Public Health* 16(1): 8-20.

Hernandez, A.V., Pasupuleti, V., Deshpande, A., Bernabe-Ortiz, A., Miranda, J.J. 2012. "Effect of rural-to-urban within-country migration on cardiovascular risk factors in low- and middle-income countries: a systematic review." *Heart* 98:185–94.

Kaczmarek, M. 2000. "A Polish version of Menopause-Specific Questionnaire, its validity and reliability." *Manuscript*

Kaczmarek, M. 2004. "The midlife well-being, gender and marital status." *Przegląd Antropologiczny – Anthropological Review* 67:57–71.

Kaczmarek, M., 2007. "The timing of natural menopause in Poland and associated factors." *Maturitas* 57: 139-153.

Kaczmarek, M. 2007a. „Określenie wieku menopauzy naturalnej w populacji polskich kobiet." *Przegląd Menopauzalny* 2:77–82.

Kaczmarek, M., Skrzypczak, M. 2012. "Perceived health status among middle-aged Polish people in relation to selected demographic and social factors." *Anthropological Review* 75(2),93-105.

Karvonen-Gutierrez, C.A. 2015 "The importance of disability as a health issue for mid-life women". *Women's Midlife Health* 1:10 DOI 10.1186/s40695-015-0011-x.

Knox, P. 1994. *Urbanization* New Jersey: Prentice Hall, Englewood Cliffs.

Knussmann, R. 1988/1992. *Anthropologie, Handbuch der Vergleichenden Biologie des Menschen* Band 1. 232-85 Fischer Verlag, Stuttgart.

Kołodziej, H., Łopuszańska, M., Jankowska, E. 2008. „Decrease in sex difference in premature mortality during system transformation in Poland." *Journal of Biosocial Science* 40(2):297-312.

Krzyżak, M., Maślach, D., Szpak, A., Piotrowska, K., Florczyk, K., Skrodzka, M., Owoc, A., Bojar, I. 2015. „Trends of potential years of life lost due to main causes of deaths in urban and rural population in Poland, 2002–2011." *Annals of Agricultural and Environmental Medicine* 22(3):564–71.

Kupiszewski, M., Durham, H., Rees, P. 1998. „Internal migration and urban change in Poland." *European Journal of Population* 14(3):265-90.

Kulu, H. 2004. "Fertility of internal migrants: comparison between Austria and Poland." *MPIDR Working Paper WP 2004-022 AUGUST 2004.*

Lasker, G.W., Mascie-Taylor, C.G.N. 1988. "The framework of migration studies." In *Biological Aspects of Human Migration* edited by C.G.N.

Mascie-Taylor and G.W. Lasker. New York, New York, Cambridge University Press, 1-19. (Cambridge Studies in Biological Anthropology Volume 2.)

Lee, J., Chia, K.S. 1993. "Estimation of prevalence rate ratios for cross sectional data: an example in occupational epidemiology [letter]." *British Journal of Industrial Medicine* 50:861-62.

Leon, D.A. 2008. "Cities, urbanization and health." *International Journal of Epidemiology* 37:4–8.

Lipowicz, A. 1999. "Fatness of children and adolescents from various socio-economic groups between 1978 and 1988." *Przegląd Antropologiczny - Anthropological Review* 62:35-40.

Lu, Y. 2010. "Rural-urban migration and health: evidence from longitudinal data in Indonesia." *Social Science and Medicine* 70(3):412–19.

Mackenbach, J.P., Stirbu, I., Roskam, A.J., Schaap, M.M., Menvielle, G., Leinsalu, M., Kunst, A.E.; European Union Working Group on Socioeconomic Inequalities in Health. 2008. "Socioeconomic inequalities in health in 22 European countries." *New England Journal of Medicine* 358(23):2468-81.

Malvezzi, M., Bertuccio P., Rosso, T., Rota M., Levi, F., La Vecchia, C., Negri, E. 2015. "European cancer mortality predictions for the year 2015: does lung cancer have the highest death rate in EU women?" *Annals of Oncology* 26:779–86.

Mann, C.J., 2003. "Observational research methods. Research design II: cohort, cross-sectional, and case-control studies." *Emergency Medicine Journal* 20:54–60.

Masterson Creber, R.M., Smeeth, L., Gilman, R.H., Miranda, J.J. 2010. "Physical activity and cardiovascular risk factors among rural and urban groups and rural-tourban migrants in Peru: a cross-sectional study." *Revista Panamericana de Salud Pública* 28:1–8.

McGranahan, G., Satterthwaite, D. 2014. "Urbanisation concepts and trends." *IIED Working Paper. IIED, London* Accessed January 4, 2016 http://pubs.iied.org/10709IIED.

McLaren, L., Hawe, P. 2005. "Ecological perspectives in health research." *Journal of Epidemiology and Community Health* 59:6–14.

Moorea, M., Gouldb, P., Kearyc, B.S. 2003. "Global urbanization and impact on health." *International Journal of Hygiene and Environmental Health* 206:269-278.

Nauman, E., VanLandingham, M., Anglewicz, P. 2016. "Migration, Urbanization and Health." In *International Handbook of Migration and*

Population Distribution edited by M.J. White International Handbook of Population 6: 451-464. DOI 10.1007/978-94-017-2_20. Springer Science+Business Media Dordrecht.

North American Menopause Society. 2010. "Overview: Introduction to menopause." *Menopause* 143(2), S4-S5.

Palloni, A., Morenoff, J. D. 2001. "Interpreting the paradoxical in the Hispanic paradox: Demographic and epidemiologic approaches." *Annals of the New York Academy of Sciences* 954(1):140-74.

Pickett, K.E., Pearl, M. 2001. "Multilevel analyses of neighbourhood socioeconomic context and health outcomes: a critical review." *Journal of Epidemiology and Community Health* 55(2):111-22.

Pumain, D. 2006. "The Urbanization Process." In *Demography: Analysis And Synthesis; A Treatise in Population Studies* edited by G.J. Wunsch, G. Caselli, J. Vallin, 319-33. Elsevier Inc.

Rėklaitienė, R., Andrijauskas, K., Bacevičienė, M. 2008. „Assessment of the quality of life among Lithuanian rural community." *Acta Medica Lituanica* 15(3):156-62.

Rogacki, H. 2007. Geografia społeczno-gospodarcza Polski. Warszawa: Wydawnictwo Naukowe PWN. 2007.

Sullivan, R., Kinra, S., Ekelund, U., Bharathi, A.V., Vaz, M., Kurpad, A. et al. 2011. "Sociodemographic patterning of physical activity across migrant groups in India: results from the Indian Migration Study." *PLoS One* 6:e24898.

Towfighi, A., Zheng, L., Ovbiagele, B. 2009. "Sex-specific trends in midlife coronary heart disease risk and prevalence." *Archives of Internal Medicine* 169:1762-1726.

Unwin, N., James, P., McLarty, D., Machybia, H., Nkulila, P., Tamin, B. et al. 2010. "Rural to urban migration and changes in cardiovascular risk factors in Tanzania: a prospective cohort study." *BMC Public Health* 10:272.

Ware, J., Sherbourne, C. 1992. "The MOS 36-item short form health survey (SF-36). I. Conceptual framework and item selection." *Medical Care* 30:473–83.

Weeks, J.R. 2008. *Population: An introduction to Concepts and Issues* (10th ed.) Wadsworth Publishing; Belmont, CA.

Welon, Z., Bielicki, T., Rogucka, E., Malina, R.M. 1999. 'Effect of education and marital status on premature mortality among urban adults in Poland, 1988-1989.' *American Journal of Human Biology* 11:397-403.

Wellons, M., Ouyang, P., Schreiner, P.J. et al. 2012. "Early menopause predicts future coronary heart disease and stroke: the Multi-Ethnic Study of Atherosclerosis." *Menopause (New York, N.Y.)* 19(10):1081-87.

White, M.J. 2015. *International Handbook of Migration and Population Distribution*. International Handbook of Population 6. DOI 10.1007/978-94-017-2_20. Springer Science+Business Media Dordrecht.

Wróblewska, W. 2002. "*Women's health status in Poland in the transition to a market economy.*" *Social Science and Medicine* 54(5):707-26.

Zagozdon, P., Kolarzyk, E., Marcinkowski, J.T. 2011. „Quality of life and rural place of residence in Polish women – population based study." *Annals of Agricultural and Environmental Medicine* 18(2):429-432.

Zatoński, W. 1996. *Evaluation of health in Poland Since 1988*. Warszawa, Polska: Centrum Onkologii, Instytut im. M. Skłodowskiej-Curie.

Żołnierczuk-Kieliszek, D., Kulik, T.B., Jarosz, M.J., Stefanowicz, A., Pacian, A., Pacian, J., Janiszewska, M., 2012. „Quality of life in peri- and post-menopausal Polish women living in Lublin Province – differences between urban and rural dwellers." *Annals of Agricultural and Environmental Medicine* 19(1), 129-133.

EEA. 2015. "The European environment — state and outlook 2015: synthesis report." European Environment Agency, Copenhagen.

EUROSTAT. 2016. "Fertility statistics" Accessed February 15, 2016 http://epp.eurostat.ec.europa.eu

GBD 2013 DALYs and HALE Collaborators. 2015. "Global, regional, and national disability-adjusted life years (DALYs) for 306 diseases and injuries and healthy life expectancy (HALE) for 188 countries, 1990–2013: quantifying the epidemiological transition." *Lancet* 386 (10009): 2145–91.

GUS. 2015. *Demographic Yearbook of Poland* Warszawa: Zakład Wydawnictw Statystycznych.

OECD 2015. *Health at a Glance 2015* OECD Indicators, OECD Publishing, Paris. Accessed February 10, 2016 http://dx.doi.org/10.1787/health_glance-2015-en.

United Nations, Department of Economic and Social Affairs, Population Division. 2008. "World Urbanization Prospects: The 2007 Revision, Highlights" (ESA/WP/205).

United Nations, Department of Economic and Social Affairs, Population Division. 2011. "Population distribution, urbanization, internal migration and development: an international perspective" (ESA/P/WP/223).

United Nations, Department of Economic and Social Affairs, World Economic and Social Survey. 2013. “Sustainable Development Challenges” (ST/ESA/ 344).

United Nations, Department of Economic and Social Affairs, Population Division. 2014. “World Urbanization Prospects: The 2014 Revision, Highlights” (ST/ESA/SER.A/352).

United Nations, Department of Economic and Social Affairs, Population Division. 2015. “World Population Prospects: The 2015 Revision, Key Findings and Advance Tables.” Working Paper No. ESA/P/WP.241.

WHO. 1981. *Research on the Menopause* WHO Technical Report Series, No. 670, Geneva: World Health Organization. 1981.

In: Biological Implications of Human Mobility ISBN: 978-1-63485-644-7
Editors: Slawomir Koziel et al.

Chapter 7

DOES SOCIAL MOBILITY OF FATHERS AFFECT AGE AT MENARCHE IN DAUGHTERS?

Slawomir Koziel and Aleksandra Gomula*
Polish Academy of Sciences,
Anthropology Unit in Wroclaw, Poland

ABSTRACT

Age at menarche is a highly ecosensitive trait, widely used as an indicator of living conditions. It is, for example, influenced by spatial migration, related usually to the improvement of the socioeconomic status of the family. There is, however, a lack of data on potential effects of social mobility on maturation, i.e., the influence of social mobility of fathers on age at menarche in their daughters. This study aimed to fill this gap. Results revealed that menarcheal age was affected neither by paternal social advancement nor social decline. Several potential comments that may be helpful in understanding these outcomes are discussed.

Keywords: social mobility, age at menarche, socioeconomic status

---

* Corresponding Author address: Polish Academy of Sciences, Anthropology Unit in Wroclaw, Podwale 75, 50-449 Wroclaw, Poland, Email: aleksandra.gomula@gmail.com.

INTRODUCTION

The age at menarche is a key-indicator of maturation in girls, commonly used in studies aimed at an assessment of living standard of a particular population (Bielicki et al. 1986, Bielicki 1989, Hulanicka et al. 2001). Research in this area is of interest to anthropologists due to the well-documented influence of socioeconomic and environmental factors responsible for significant differences between populations and social groups.

Age at menarche is strongly determined by genetic factors (Eveleth and Tanner 1990, Sanches-Andreas 1997). However, at the same time it is also modified by the general living conditions, such as: quality of food, health status, physical activity, psychosomatic stress level, number of siblings, birth order, as well as parental occupation or educational and income level (Artaria and Henneberg 2000, Hulanicka et al. 1990, Hulanicka et al. 2001, Łaska-Mierzejewska and Olszewska 2004, Malina et al. 1997, Salces et al. 2001, Szwed et al. 2013, Ulijaszek et al. 1991). Many researchers also indicate the important influence of the level of urbanization (Hulanicka et al. 1999), living conditions (Biswas et al. 2013) as well as parental spatial migration into a new living environment (Gomula and Koziel 2015).

The most important outcome from a sociological point of view, is the influence of socioeconomic status (SES) on the variation in age at menarche in girls from both developing and developed countries (Deardorff et al. 2014, Eveleth and Tanner 1990, James-Todd et al. 2010, Liestol and Rosenberg 1995). Previous studies on the relation between SES and age at menarche revealed ambiguous results: according to some of them a low SES was related to early menarche (James-Todd et al. 2010), while others found no relationship (Buttke et al. 2012), or provided inconsistent results, depending on the used indicator (Windham et al. 2004, Lee et al. 2007). In Poland during the last 40 years, most of the research pointed at a significant social variation in age at menarche (Hulanicka et al. 1990, Hulanicka et al. 2001, Łaska-Mierzejewska and Olszewska 2004, Szwed et al. 2013).

In most of the modern postindustrial societies, where significant social inequality is observed, the phenomenon of social mobility is taking place. It concerns the changes in socioeconomic status of an individual in comparison to the starting position determined by parental socio-economic status. Upward social mobility, denoting a direction higher than parents' SES, is related to an individual's body height. It can be considered in the context of selective social mobility: taller individuals have a higher probability to gain a better SES during their lifespan than shorter ones, in comparison to the starting position determined by the socio-economic status of parents. This phenomenon was found in many studies conducted on both Polish (Bielicki and Charzewski 1983, Bielicki and Szklarska 2000, Bielicki and Waliszko 1992, Szklarska et al. 2007) as well as western samples (Olivier 1977, Schreider 1967, Schumacher and Knussman 1979, Tanner 1966).

While all previous studies are related to the biological effect of upward social mobility with respect to the previous generation, i.e., in comparison to the position on the scale of education or occupation of fathers or parents of subjects, there is a lack of data on the potential long-term effects of social mobility on the development of subsequent generations, i.e., on the maturation of the second generation that has not yet reached a final position on the social ladder: the influence of social mobility of fathers on age at menarche in their daughters.

The aim of this study was to assess the potential effect of social mobility of fathers, estimated by their educational level compared to parental grandfathers, on age at menarche in daughters, while controlling for the socio-economic status of the family.

METHODS

Data was collected during 5th Polish Anthropological Survey carried out between 2012 and 2014 in three cities (Warsaw, Wroclaw, Lodz), four towns (Bystrzyca Klodzka, Pinczow, Wolsztyn, Siemiatycze) and rural areas belonging to the districts of these towns. The project was funded by the National Science Centre in 2011-2014 under the grant number: NN303804540.

The study included schoolchildren and adolescents aged 7 to 19 years. Research was conducted in randomly selected primary and secondary schools in cities and most of the schools in towns and villages. Only girls were involved in this study. Information on age at menarche was obtained by using two methods: *status quo* and retrospective. Since some research revealed different results acquired by these two methods (e.g., Artaria and Henneberg 2000), the use of both aimed to verify their convergence.

The *status quo* method was based on the answers of girls ("yes" or "no") on the question: "Have you already had your first menstruation?". This method involved 2692 participants aged 7-18.5 years. The mean age at menarche was assessed using probit or logit analyses. When the assumption of goodness-of-fit was met (p-value for the Chi-square test ($\chi 2$) was higher than 0.05) the probit model was used, while where it was not met, the logit model was run

(Finney 1952). The results obtained by both methods are very similar, and even considered as indistinguishable (Vincent 2010). Whenever the significance of differences between more than two means was tested, the Bonferroni correction was used. Analyses were conducted using StatsDirect.

The retrospective method included only girls older than 16 years of age who had already had their menarche and provided the date of its occurrence (N = 624). Age at menarche was calculated based on the difference between the date of the first menstrual period and the date of birth. Whenever the received information on menarche was limited to the month and year, the middle of the month was used. The following statistical tests were run with respect to these data: one-way and two-way ANOVA as well as HSD Tukey *post-hoc* and Fisher's LSD *post-hoc* tests. All of them were conducted with the use of Statistica 10.0. software (StatSoft 2010).

Moreover, parents provided information on the socio-economic status of the family. They also gave written consent to their children's participation in this study.

Three categories of social mobility were determined by comparing the level of education of the father and paternal grandfather. The education level was divided into 4 categories: elementary, trade, secondary schools and university degree. Whenever the father gained a level of education higher than the grandfather, it was regarded as upward social mobility, in case he had a lower education level than the grandfather – his social mobility was downward, when there was no change in education – it was coded as no changes in social mobility.

The assessment of general socioeconomic status (SES) was obtained by the use of factor analysis with the principal components method (PCA) (concerning location size, family size – number of siblings, father's and mother's education). Scores of the first component were included into further analyses as a proxy for SES. Three groups of SES (low, medium and high) were based on the tertiles. Results of PCA are presented in Table 1.

RESULTS

The results of goodness of fit are presented in Tables 2 and 3. Descriptive statistics for age at menarche obtained by the *status quo* method in particular social mobility groups, as well as in the groups included in SES categories are provided, respectively, in Tables 4 and 5. Descriptive statistics for age at menarche obtained by retrospective method in particular social mobility groups, as well as in the groups included in SES categories are provided, respectively, in Tables 6 and 7.

No significant differences in age at menarche assessed by *status quo* methods were revealed between three categories of social mobility ($p > 0.05$) (Figure 1). These results remained insignificant also after including families' SES ($p > 0.05$) (Figure 2).

Table 1. Results of PCA for four social variables

social variables	loadings
father's education level	0.74
mother's education level	0.66
urbanization	0.84
family size	0.85
eigen value (variation)	2.41 (60.21%)

Table 2. Goodness-of-fit and number of iterations for age at menarche in subsequent social mobility categories (status quo method)

social mobility	$\chi 2$	p	df	iterations
upward	18.19	0.395	22	6
no changes	21.78	0.473	22	7
downward	8.70	0.995	22	7

Table 3. Goodness-of-fit and number of iterations for age at menarche in subsequent SES groups and social mobility categories (status quo method)

SES	social mobility	$\chi 2$	p	df	iterations
high	upward	24.44	0.33	22	9
	no changes	5.46	> 0.99	22	10
	downward	10.56	0.98	22	14
medium	upward	14.87	0.87	22	9
	no changes	20.38	0.50	21	11
	downward	3.34	> 0.99	22	13
low	upward	3.30	> 0.99	21	11
	no changes	11.40	> 0.97	22	12
	downward	10.65	0.98	22	12

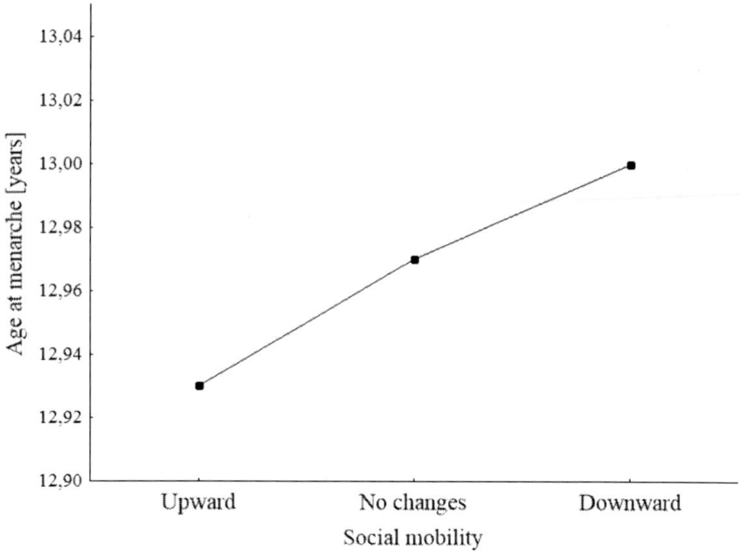

Figure 1. Age at menarche in subsequent social mobility groups (*status quo* method)

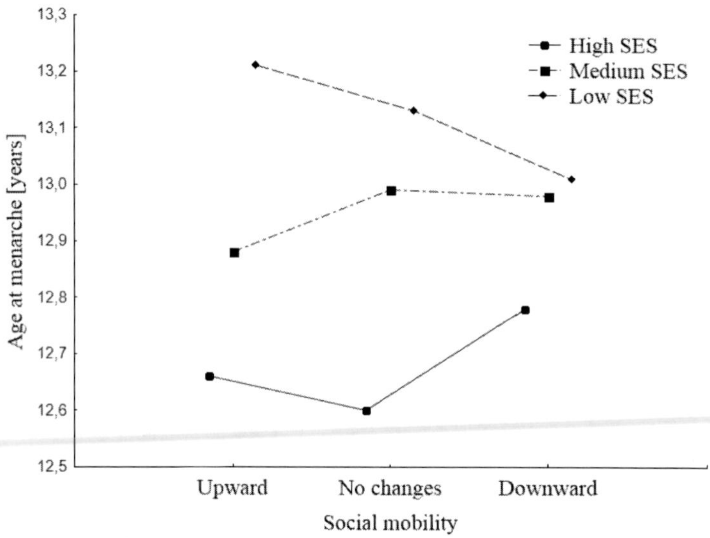

Figure 2. Age at menarche in subsequent social mobility groups depending on SES category (*status quo* method)

Table 4. Descriptive statistics of age at menarche in particular social mobility groups (status quo method)

social mobility	N	mean	95% CI	p90	SD	SE
upward	1439	12.93	12.77- 13.10	14.33	1.09	0.03
no changes	1042	12.94	12.77-13.13	14.16	0.95	0.03
downward	211	13.00	12.63-13.40	14.48	1.16	0.08

N – number of participants, 95% CI - 95% confidence intervals, p90 - 90th percentiles, SD – standard deviation, SE – standard error.

Table 5. Descriptive statistics of age at menarche in particular SES and social mobility groups (status quo method)

SES	social mobility	N	Mean	95% CI	p90	SD	SE
high	upward	507	12.66	12.37-12.96	14.53	1.46	0.06
	no changes	280	12.60	12.26-12.97	13.84	0.97	0.06
	downward	30	12.78	11.51-14.58	14.27	1.16	0.21
medium	upward	556	12.88	12.64-13.13	14.29	1.10	0.05
	no changes	335	12.99	12.67-13.31	14.25	0.98	0.05
	downward	79	12.98	12.53-13.65	14.31	1.04	0.12
low	upward	332	13.21	12.89-13.52	14.55	1.05	0.06
	no changes	392	13.13	12.83-13.44	14.65	1.19	0.06
	downward	101	13.01	12.45-13.65	14.83	1.42	0.14

N – number of participants, 95% CI - 95% confidence intervals, p90 - 90th percentiles, SD – standard deviation, SE – standard error.

Table 6. Descriptive statistics of age at menarche in particular social mobility groups (retrospective method)

social mobility	N	Mean	95% CI	SD	SE
upward	363	12.67	12.54-12.81	1.27	0.07
no changes	231	12.64	12.47-12.81	1.31	0.09
downward	30	12.57	11.96-13.18	1.63	0.30

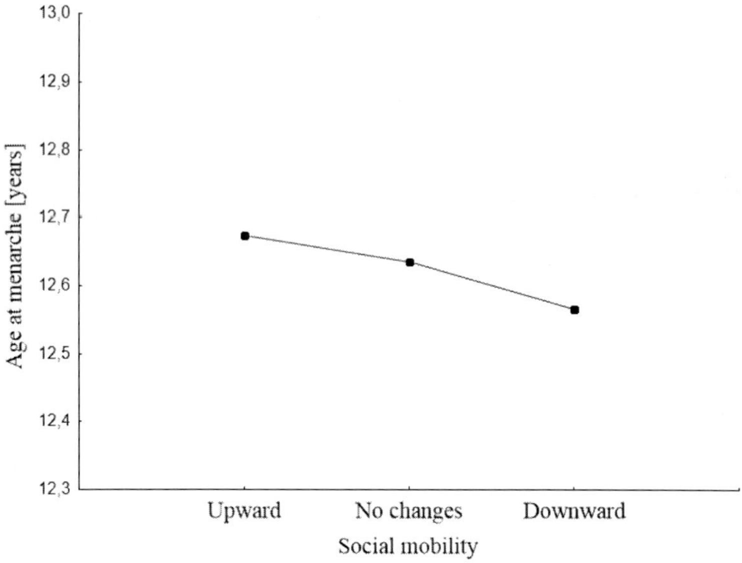

Figure 3. Age at menarche in subsequent social mobility groups (retrospective method)

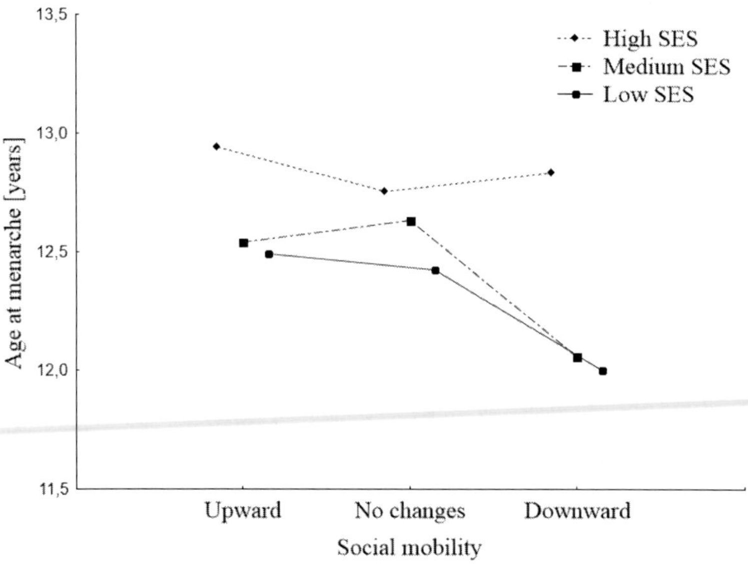

Figure 4. Age at menarche in subsequent social mobility groups depending on SES category (retrospective method)

Table 7. Descriptive statistics of age at menarche in particular SES and social mobility groups (retrospective method)

SES	social mobility	N	mean	95% CI	SD	SE
High	upward	80	12.49	12.22-12.76	1.21	0.14
	no changes	44	12.42	12.09-12.76	1.10	0.17
	downward	2	12.00	-	2.79	1.97
medium	upward	142	12.54	12.33-12.76	1.30	0.11
	no changes	82	12.63	12.35-12.91	1.28	0.14
	downward	9	12.06	10.47-13.65	2.07	0.69
Low	upward	133	12.94	12.73-13.15	1.24	0.12
	no changes	98	12.75	12.47-13.04	1.43	0.15
	downward	18	12.83	12.17-13.50	1.34	0.32

N – number of participants, 95% CI - 95% confidence intervals, SD – standard deviation, SE – standard error

Similar results were obtained with respect to the retrospective method. One-way ANOVA revealed no significant differences in age at menarche between three groups of social mobility ($F(2, 621) = 0.13$, $p > 0.05$) (Figure 3). To exclude the possibility of a disturbing effect of the different mean age values of girls, one way ANOVA was run, indicating that this feature did not differ significantly with respect to social mobility groups ($F(2, 621) = 0.82$, $p > 0.05$). There were also no differences in age at menarche between the three groups within SES categories ($F(4, 599) = 0.53$, $p > 0.05$). However, although the effect of interaction between factors was insignificant, the net effect of SES was significant ($F = 3.10$, $p < 0.05$), indicating that SES differentiated age at menarche of girls whose fathers have experienced social advancement (Fisher's LSD *post-hoc* test: $p < 0.05$). A graphic presentation of those effects are shown in Figure 4.

DISCUSSION

Results of this study revealed that social mobility indicated by the educational level of fathers compared to paternal grandfathers did not affect age at menarche of daughters irrespective of the method used. Although it could be suspected that social advancement would accelerate maturation, as it was often found in terms of increased height, this result seems to be explainable. There are several potential comments that may be helpful in understanding these outcomes.

One of them could be a different effect of social mobility depending on the offspring's sex. For example, research focused on the social mobility of children in relation to fathers revealed significant differences in height between downwardly and upwardly mobile boys, but not girls (e.g., Cernerud 1995). This sex-dependent differences might be caused by different vulnerability to environmental conditions, as boys are more ecosensitive, compared to girls (see also: Bielicki and Charzewski 1983).

It is, however, also possible that this biological effect of social advancement is only maintained in the first generation, while in the second one it may be covered by the influence of some different factors. This could be especially true regarding age at menarche, due to its high sensitivity to environmental conditions. It reacts more intensively on changes within individual socioeconomic context than for example height or weight (see: Koziel et al. 2014). Therefore, this biological trait may be under a stronger influence of the very present living standards (than e.g., height which is rather dependent on the long-term impact of the environment), reacting to these conditions more rapidly and eliminating the effect of previous, less intensive events, such as social advancement of parents.

Another issue important to mention refers to the relationship between age at menarche and nutritional status (e.g., Freedman et al. 2002, Ziomkiewicz and Koziel 2015) which, in turn, is under the influence of socioeconomic conditions (e.g., Lipowicz 1999, Wang 2001). This last relationship, however, may have an opposite direction depending on the level of development of a country. In developing countries, children from higher SES strata have more often extended body weight than those from low SES families, while in developed ones this trend is reversed: when parents have high SES, their children are less overweight or obese than their peers from low SES groups. The reason of this phenomenon is caused by an awareness of the importance of healthy eating habits and lifestyle in better educated families from developed countries, comparing to people with low SES, who eat more cheaply, but unhealthy and fattening food. In contrast, in developing countries, higher SES families have better access to food, but no awareness of the risk following obesity, while those with low SES are not able to secure such an amount of food, which also often results in undernutrition. Since there is a relationship between body mass and maturation in girls, consequent different patterns are observed with respect to menarche (Braithwaite et al. 2009, Deardorff et al. 2014, Parent et al. 2003). Poland seems now to be on the border between a developed and developing country. Therefore, the impact of education on healthy lifestyle may not be as visible and unequivocal. In this sample of girls, BMI values standardised to age were also not significantly different depending on paternal education, although a slight trend toward developed countries' pattern was observed (Nowak-Szczepanska and Gomula 2014). This second fact concerning the place of the Polish population in the area "between" two categories of countries with varying status of life, may contribute to the lack of noticeable relationship between social advancement and maturation in girls.

The fact of using education level as an indicator of social mobility could also influence the obtained results, since this feature, however, not necessarily directly reflects the real family standard of living: the relationship between education and socioeconomic status of the family may not be linear. In fact, depending on the situation on the labour market, better paid workers may be those with lower education, but also with some specific training. Furthermore, in the capitalist system, unemployment is a universal phenomenon, affecting people regardless of their level of education. Age at menarche is considered as an indicator of living conditions, which was also confirmed in this sample of Polish schoolgirls. Indeed, general socio-economic status of the family significantly differentiated maturation ($p < 0.05$), however, there were no significant differences in menarcheal age between two extreme groups of paternal education level ($p > 0.05$) (see also: Durda 2011, Papadimitriou et al. 2008). This remark may contribute to the validity of the assumption that in the modern Polish population, the level of paternal education may not be directly related to the living conditions of the family and, following on this, social mobility based on educational level may not reflect the mobility in terms of social advancement. This phenomenon could be conducive to some extent, in the explanation of the lack of relationship between social mobility and girls' maturation.

This study also has some limitations important to mention, as they could also affect the obtained results. Over more than the last half a century in Poland, the situation regarding access and, consequently, level of education has changed radically. Grandfathers of subjects were born around the time of the Second World War. In the early post-war years, a better level of education was very difficult to obtain, while the fathers of subjects were born and lived in more favourable conditions enabling them to get a secondary education (in 1966 nearly 64% of fathers had only primary education, while in 2012 it was not even 6% (Gomula and Nowak-Szczepanska (2014)). Therefore, it was rather rare to gain a lower level of education by fathers compared to grandparents. Consequently, the number of girls in the downwardly mobile fathers' groups was very low and this could, in turn, affect or even skew obtained results.

On the other hand, considering these results in the perspective of selective social mobility, it seems that sexual maturity, especially in females, has a different meaning with respect to this issue than e.g., body height. The upward selective social mobility of taller people (particularly men) compared to shorter ones has been widely documented. It was explained by the genetic link between tallness and some psychological traits which facilitate educational advancement, as well as by the significance of tall stature in the climbing the social ladder (Bielicki and Charzewski 1983). From this point of view, early age at menarche seems not to be beneficial in gaining higher social status. This is especially due to its relationship with early reproduction: girls who reach their sexual maturity earlier, compared to those who mature later, also have their first sexual intercourse and pregnancy earlier (Udry 1979). In this context, early puberty followed by early motherhood does not facilitate social advancement, making it even more difficult, since for these women, in fact, getting an education is hampered as they have to dedicate themselves to child rearing.

Research on social mobility and body height as well as weight, mostly indicates that upwardly mobile people are taller compared to those whose mobility had a downward direction, while (in developed countries) social advancement is related to lower body weight. These two biological features are commonly considered as health predictors. Taken from a perspective of healthy well-being, in the case of menarche, this issue is more ambiguous. According to the life history theory, females may reach sexual maturity when the environmental circumstances are good enough for pregnancy and raising her offspring (e.g., Ellison 2003). Hence, the better living conditions are, the earlier the menarche may occur. On the other hand however, early menarche is related to many serious diseases, such as cardiovascular diseases, metabolic syndromes or even cancers (Allison and Hyde 2013, Frontini et al. 2003). Therefore, in the long run, it is not advantageous for females to mature in a very young age. In the Polish population, girls already have gained a mean age at menarche similar to those observed in western countries (e.g., in Germany (Gohlke and Woelfle 2009), Sweden (Gustafsson and Hammarström 2012), Netherlands (Talma et al. 2013), USA (Carwile et al. 2015), or Canada (Al-Sahab et al. 2010)), thus, further acceleration of maturation in socio-economically privileged group would not be beneficial as it could result in enhanced morbidity. Therefore, early age at menarche should not be merely considered within the context of socioeconomic privileges and as an indicator of better living conditions, as with regard to increased body height, but while interpreting results also its adverse impact on future health conditions should be underlined.

REFERENCES

Al-Sahab, Ban, Ardern, Chris I., Hamadeh, Mazen J., and Hala Tamim. 2010. "Age at menarche in Canada: results from the National Longitudinal Survey of Children and Youth." *BMC Public Health* 10(1):736.

Allison, Carlie M., and Janet S. Hyde. 2013. "Early menarche: confluence of biological and contextual factors." *Sex Roles* 68(1-2):55-64.

Artaria, Myrtati D., and Maciej Henneberg. 2000. "Why did they lie? Socio-economic bias in reporting menarcheal age." *Annals of Human Biology* 27(6):561-569.

Bielicki, Tadeusz, and Janusz Charzewski. 1983. "Body height and upward social mobility." *Annals of Human Biology* 10(5):403-408.

Bielicki, Tadeusz., and Hanna Waliszko. 1992. „Stature, upward social mobility and the nature of statural differences between social classes." *Annals of Human Biology* 19:589-593.

Bielicki, Tadeusz., and Alicja Szklarska. 2000. „Are social-class differences in stature party genetic? A hypothesis revised." *American Journal of Human Biology* 12:97-101.

Braithwaite, Dejana, Moore, Dan H., Lustig, Robert H., Epel, Elissa S., Ong, Ken K., Rehkopf, David H., Wang, May C., Miller, Suzanne M., and Robert A. Hiatt. 2009. "Socioeconomic status in relation to early *menarche* among black and white girls." *Cancer Causes and Control* 20(5):713-720.

Buttke, Danielle E., Sircar, Kanta, and Colleen Martin. 2012. "Exposures to endocrine-disrupting chemicals and age at menarche in adolescent girls in NHANES (2003-2008)." *Environmental Health Perspectives* 120:1613e8.

Carwile, Jenny L., Willett, Walter C., Spiegelman, Donna, Hertzmark, Ellen, Rich-Edwards, Janet, Frazier, Lindsay A., and Karin B. Michels. 2015. "Sugar-sweetened beverage consumption and age at menarche in a prospective study of US girls." *Human Reproduction* 30(3):675-683.

Cernerud, Lars. 1995. "Height and Social Mobility A Study of the Height of 10 Year Olds in Relation to Socio-economic Background and Type of Formal Schooling." *Scandinavian Journal of Public Health* 23(1):28-31.

Deardorff, Julianna, Abrams, Barbara, Ekwaru John P., and David H. Rehkopf. 2014. "Socioeconomic status and age at *menarche*: an examination of multiple indicators in an ethnically diverse cohort." *Annals of Epidemiology* 24(10):727-733.

Durda, Magdalena. 2011. "Biological status of adolescents in relations to their lifestyle behaviours and family's socioeconomic status." In *Health and well-being in adolescence. Part one Physical Health and subjective well-being*, edited by Maria Kaczmarek, 111-137. Poznan: Bogucki Wydawnictwo Naukowe.

Ellison, Peter T. 2003. "Energetics and reproductive effort." *American Journal of Human Biology* 15(3):342-351.

Freedman, David S., Khan, Laura K., Serdula, Mary K., Dietz, William H., Srinivasan, Sathanur R., and Gerald S. Berenson. 2002. "Relation to age at *menarche* to race, time period and anthropometric dimensions: the Bogalusa heart study." *Pediatrics* 110(4):e43.

Frontini, Maria G., Srinivasan, Sathanur R., and Gerald S. Berenson. 2003. "Longitudinal changes in risk variables underlying metabolic Syndrome X from childhood to young adulthood in female subjects with a history of early *menarche*: the Bogalusa Heart Study." *International Journal of Obesity Related Metabolic Disorders* 27(11):1398-1404.

Gohlke, Bettina, and Joachim Woelfle. 2009. "Growth and puberty in German children." *Deutsches Ärzteblatt International* 106(23):377-382.

Gomula, Aleksandra, and Natalia Nowak-Szczepanska. 2014. "Polish Anthropological Nationwide Field Investigation." In *Anthropological surveys of Polish schoolchildren, 1966-2012. Secular changes and social variation*, edited by Slawomir Koziel, Natalia Nowak-Szczepanska, and Aleksandra Gomula, 21-36. Wroclaw: Publishing house Arboretum.

Gustafsson, Per E., and Anne Hammarström. 2012. "Socioeconomic disadvantage in adolescent women and metabolic syndrome in mid-adulthood: an examination of pathways of embodiment in the Northern Swedish Cohort." *Social Science and Medicine* 74(10):1630-1638.

Hulanicka, Barbara, Gronkiewicz, Lidia, and Jan Koniarek. 2001. „Effect of familial distress on growth and maturation of girls: A longitudinal study." *American Journal of Human Biology* 13:771-776.

Hulanicka, Barbara., Brajczewski, Czesław., Jedlińska, Wanda., Sławińska, Teresa, and Anna Waliszko. 1990. *City – Town – Village. Growth of children in Poland in 1988*. Wroclaw: Monographies of the Institute of Anthropology Polish Academy of Sciences.

James-Todd, Tamarra, Tehranifar, Parisa, Rich-Edwards, Janet, Titievsky, Lina, and Mary B. Terry. 2010. "The impact of socioeconomic status across early life on age at menarche among a racially diverse population of girls." *Annals of Epidemiology* 20:836-842.

Koziel, Slawomir, Nowak-Szczepanska, Natalia, and Aleksandra Gomula. 2014. *Anthropological surveys of Polish schoolchildren, 1966-2012. Secular changes and social variation.* Wroclaw: Publishing house Arboretum.

Lee, Joyce M., Appugliese, Danielle, Kaciroti, Niko, Corwyn, Robert F., Bradley, Robert H., and Julie C. Lumeng. 2007. "Weight status in young girls and the onset of puberty." *Pediatrics* 119:624330.

Lipowicz, Anna. 1999. "Fatness of children and adolescents from various socio-economic groups between 1978 and 1988." *Anthropological Review* 62:35–40.

Łaska-Mierzejewska, Teresa, and Elżbieta Olszewska. 2004. "The maturation rate of girls living in rich and poor rural regions of Poland before and after the transformation of 1989." *HOMO-Journal of Comparative Human Biology* 55(1):129-142.

Papadimitriou, Anastasios, Fytanidis, Grigorios, Douros, Konstandinos, Bakoula, Chryssa, Nicolaidou, Polyxeni, and Andreas Fretzayas. 2008. "Age at *menarche* in contemporary Greek girls: evidence for levelling-off of the secular trend." *Acta Pædiatrica* 97(6):812-815.

Nowak-Szczepanska, Natalia, and Aleksandra Gomula. 2014. Changing pattern of social variation in height and BMI between 1966 and 2012 in Polish schoolchildren. In *Anthropological surveys of Polish schoolchildren, 1966-2012. Secular changes and social variation*, edited by Slawomir Koziel, Natalia Nowak-Szczepanska, and Aleksandra Gomula, 61-85. Wroclaw: Publishing house Arboretum.

Olivier, Georges. 1977. "Hypothese sur la cause des differences biologiques entre las categories socials." *L'Anthropologie* 81:579-600.

Parent, Anne S., Teilmann, Grete J., Juul, Anders, Skakkebaekn, Niels E., Toppari, Jorma, and Jean-Pierre Bourguignon. 2003. "The timing of normal puberty and the age limits of sexual precocity: variations around the world, secular trends, and changes after migration." *Endocrine Reviews* 24(5):668-693.

Schreider, Eugene. 1967. "Possible selective mechanisms of social differentiation in biological traits." *Human Biology* 39:14-20.

Schumacher, Astrid, and Renate Knussman. 1979. "Are the differences in stature between social classes a modification or an assortment effect?" *Human Biology* 8:809-812.

StatSoft. 2010. *STATISTICA (Data Analysis Software System)*, Version 10. www.statsoft.com

Szklarska, Alicja, Kozieł, Sławomir, Bielicki, Tadeusz, and Robert M. Malina. 2007. „Influences of height on attained level of education in males at 19 years of age." *Journal of Biosocial Sciences* 39:575-582.

Szwed, Anita, John, Aleksandra, Czapla, Zbigniew, and Magdalena Kosińska. 2013. „Influence of socioeconomic factors on age at menarche of Polish girls." *Anthropologischer Anzeiger* 70(4):455-470.

Talma, Henk, Schönbeck, Yvonne, van Dommelen, Paula, Bakker, Boudewijn, van Buuren, Stef, and Remy A. HiraSing. 2013. "Trends in menarcheal age between 1955 and 2009 in the Netherlands." *PLoS One* 8(4):e60056.

Tanner, James M. 1966. "Galtonian eugenics and the study of growth." *Eugenics Review* 58:122-135.

Udry, Richard J. 1979. "Age at menarche, at first intercourse, and at first pregnancy." *Journal of Biosocial Science* 11(04):433-441.

Vincent, Kim. 2010. "Probit analysis." The San Francisco State University. Dpt of Biology: userwww.sfsu.edu/efc/classes/biol710/probit/ProbitAnalysis.pdf.

Wang, Youfa. 2001. "Cross-national comparison of childhood obesity: the epidemic and the relationship between obesity and socioeconomic status." *International Journal of Epidemiology* 30(5):1129-1136.

Windham, Gayle C., Bottomley, Christian, Birner, Cecilie, and Laura Fenster. 2004. "Age at menarche in relation to maternal use of tabacco, alcohol, coffee, and tea during pregnancy." *American Journal of Epidemiology* 159:862e71.

Ziomkiewicz, Anna, and Sławomir Kozieł. 2015. "Grow first, gain fat in the meantime. Longitudinal study of anthropometric changes around *menarche*." *Anthropological Review* 78(2):169-181.

In: Biological Implications of Human Mobility ISBN: 978-1-63485-644-7
Editors: Slawomir Koziel et al.

Chapter 8

IMPACT OF REGIONAL MIGRATION AND SOCIAL MOBILITY ON VARIATION IN ADULT HEIGHT, WEIGHT AND BODY MASS INDEX: EVIDENCE FROM A BRITISH COHORT STUDY

Monika Krzyżanowska[1,*]
and C. G. Nicholas Mascie-Taylor[2]
[1]Department of Human Biology, University of Wrocław, Poland
[2]Department of Archaeology and Anthropology, University of Cambridge, Cambridge, UK

ABSTRACT

Using a sample of the British National Child Development Study (NCDS), the regional variation, intra- and inter-generational migration and social mobility (migration between social classes) as well as the relationships between social and geographical intra- and inter-generational mobility within Britain were examined in relation to height, weight and Body Mass Index (BMI). The relationship between inter-generational social mobility and biosocial variables was also analysed. The extent of geographical migration was determined and categorised as none (no change in place of residence), in-migrants (in-migration into

* Corresponding Author address: Department of Human Biology, University of Wrocław, Kuźnicza 35, 50-138 Wrocław, Poland; Email: monika.krzyzanowska@uwr.edu.pl.

another region) and out-migrants (emigration from one region to another). The net effect of migration (the difference between in- and out-migrants) was also calculated. Social mobility was categorised as none (no change in social class), upwardly mobile (moving up one or more social classes) or downwardly mobile (going down one or more social classes). The results confirmed a selective migration in relation to height and BMI. It appeared that people who were geographically or socially upwardly mobile within Britain in the second half of the twentieth century tended to be taller and have lower BMI than non-migrants or non-mobile ones. Social mobility and geographical migration were acting in a simple additive way. The most significant predictor of social mobility in both sexes was education of the cohort member, followed by social class of the father and education of the father, which indicates that social mobility in Britain takes place largely on meritocratic principles.

Keywords: height, weight, BMI, migration, intra- and inter-generational social mobility

INTRODUCTION

Throughout human evolution, individuals have tended to group together in clusters in many different types of environments. The spatial separation of clusters tends to act as a barrier and mating is more likely to occur within, than between, clusters. There are many studies that have examined the extent of geographical separation on variation in human biological characteristics: for example, the classic work on genetic polymorphisms by Mourant et al. (1976).

The effects of geographical migration have been studied for over 100 years beginning with the works of Fishberg and Boas (quoted by Lasker and Mascie-Taylor, 1988).Migration impacts on both recipient and donor populations and movement may be selective or random (Mascie-Taylor and Little, 2004).Selective migration promotes or maintains differences between groups or regions (maintains heterogeneity) whereas random migration reduces differences (promotes homogeneity). Mascie-Taylor (1985) showed that migration for IQ is selective. Mascie-Taylor and Lasker (1987) modelled the regional migration in Britain for ABO and Rh blood groups and showed that the spatial clines would disappear in about ten generations.

In general geographical migration studies show that body height is higher in people who change their place of residence (Kobyliansky and Arensburg, 1977; Kim, 1982; Wheeler and Tan, 1983; Bernis, 1984; Steegmann, 1985; Singh and Harrison, 1996). However, some researchers have not found a connection between body height and migration (Lasker, 1954; Malina et al., 1982; Zielińska, 1991). Migrants have also been shown to have more advanced sexual maturation and higher body mass index (BMI) than non-migrants (Garnier et al., 2003; Nagel et al., 2009).

Besides geographical separation, clusters are often stratified into socioeconomic groupings, so human populations can be differentiated both horizontally (geographical) and vertically (socially) (Harrison and Boyce, 1972). Differences between social groupsin mean IQ scores(Burt, 1961; Gibson, 1970; Waller, 1971; Mascie-Taylor and Gibson, 1978; Nettle, 2003) and genetic markers(Cartwright et al., 1978) have also been reported.

Social class differences in height, weight and body mass index (BMI) have been widely reported in many countries in both children and adults (Mascie-Taylor, 1984; Lasker and Mascie-Taylor, 1996; Komlos and Kriwy, 2003; Heineck, 2006; Batty et al., 2009; Cardoso and Caninas, 2010; Kues, 2010; Singh-Manoux et al., 2010). There is a general tendency for mean height and weight to decline from higher to lower social classes while BMI shows the opposite trend (Lasker and Mascie-Taylor, 1989; Li et al., 2004). If movement between social classes (social mobility) was at random, then the mean differences between classes would tend to disappear depending on the length of time the mobility is in effect and the number of people moving.

However, if migration is selective, then social class differences would be maintained to a greater or lesser extent. In general social mobility studies show that upwardly socially mobile individuals are, on average, taller and heavier than the non-socially mobile, while the downwardly socially mobile are, on average, shorter (Mascie-Taylor, 1984; Cernerud, 1995; Bielicki and Szklarska, 2000; Hart et al., 2008). A similar finding has been reported for IQ and educational attainment (Gibson and Mascie-Taylor, 1973; Mascie-Taylor and Gibson, 1978; Gibson et al., 1983; Deary et al., 2005; Johnsonet al., 2010).

This chapter examines the extent of geographical variation and regional migration as well as social class differences and social mobility in relation to height, weight and Body Mass Index using a British Cohort Study.

The National Child Development Study

The National Cohort Study (NCDS) is a longitudinal study of all children born in England, Wales and Scotland in the week 3–9th March 1958. The children and their families were periodically re-studied in 1965, 1969 and

1974. Thereafter the index child was followed up into adulthood, and there were re-studies when the child was 33 and 40 years of age.

Figure 1. Map of Britain showing the nine standard regions. Y + H = Yorkshire and Humberside, East M = East Midlands, West M = West Midlands and NW = North West.

Families were placed into one of the 9 standard regions comprising North, Yorkshire and Humberside, East Midlands, East and South East, South West, West Midlands, North West, Wales, Scotland (see Figure 1). The extent of geographical migration was determined and categorised as none (no change in place of residence), in-migrants (in-migration into another region) and out-migrants (emigration from one region to another).

Social class was defined by the Registrar General's 5-point occupational scale, where I refers to professional, II to intermediate (mainly managerial), III skilled worker, IV semi-skilled worker and V unskilled. Social mobility was categorised as none (no change in social class), upwardly mobile (moving up one or more social classes) or downwardly mobile (going down one or more social classes). In order to differentiate between social and geographical movement, movement between social classes is referred to as social mobility, while movement between regions is referred to as geographical or regional migration.

1. Geographical Variation and Regional Migration in Relation to Height, Weight and Body Mass Index

1.1.Geographical Variation

Information was available on 2090 father and son pairs and the extent of intra-generational regional migration was determined between 1958 and 1974 for fathers and between 1991 and 2000 for sons (the index child born in 1958).Table 1 presents the percentages in each region and about a third of fathers and sons lived in the East and South-East region and the least in Wales (just over 5%).

Highly significant regional variation was found in fathers mean height, weight and BMI but not for sons. Figures 2 to 4 present the means at the four surveys conducted between 1958 and 1974. As Figure 2 shows the tallest fathers were, on average, living in the East and South-East region and the shortest in Wales with a mean difference of about 3cms.The regions also roughly fell into three groups, the tallest group comprising West Midlands, East Midlands and East and South-East; an intermediate group of the North West, North, Yorkshire and Humberside and the South West, and the shortest group made up of Scotland and Wales.

For weight the highest mean was in the West Midlands (about 77kg) and the lowest in the South-West (about 73.5kg) with an overall range of 3.5kg.Fathers living in Yorkshire showed an average increase in weight of about 0.5kg between 1958 and 1974 but for most other regions there was little intra-regional change in mean.

Table 1. Percentages of inhabitants in the 9 regions (Krzyżanowska and Mascie-Taylor, 2011b)

Regions	Fathers		Sons	
	1958	1974	1991	2000
	%	%	%	%
North	8.3	8.3	6.7	7.0
Yorkshire +Humberside	7.8	7.2	7.9	7.5
East Midlands	7.2	7.5	7.5	7.6
East + South East	33.0	33.2	34.6	33.3
South West	6.2	7.1	8.8	9.5
West Midlands	10.2	9.7	9.4	9.5
North West	10.4	10.1	9.2	9.4
Wales	5.2	5.5	5.5	5.4
Scotland	11.7	11.4	10.4	10.8
Total	100	100	100	100

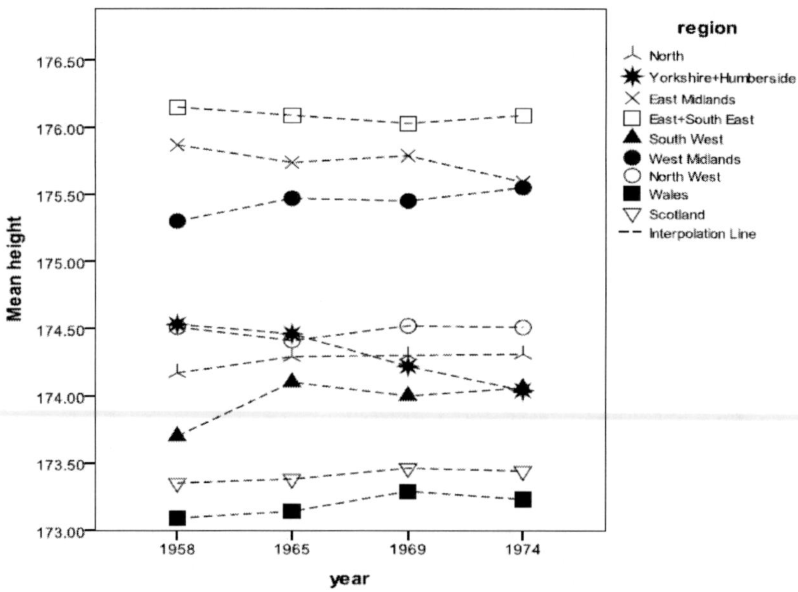

Figure 2. Fathers' mean height by region (Krzyżanowska and Mascie-Taylor, 2011b).

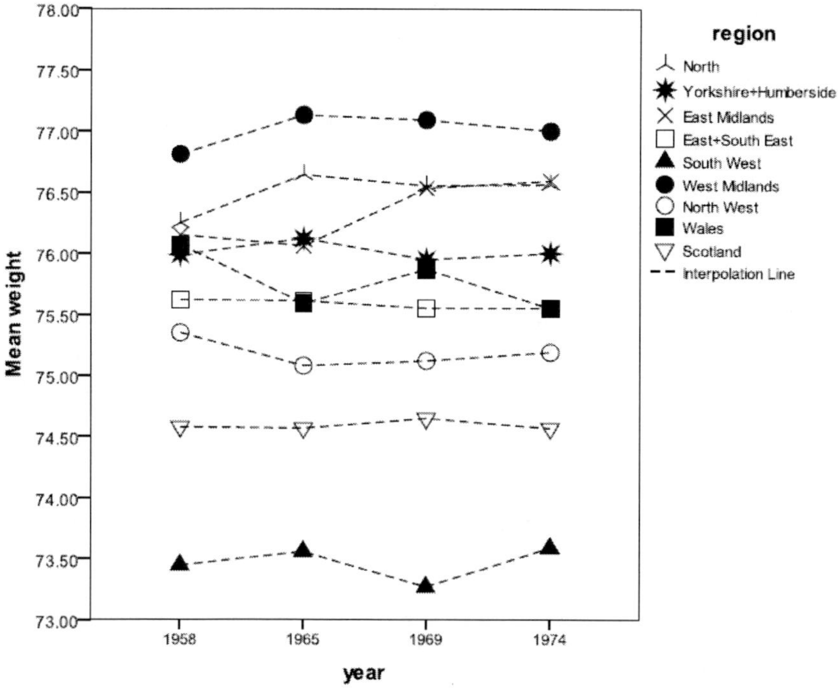

Figure 3. Fathers' mean weight by region (Krzyżanowska and Mascie-Taylor, 2011b).

Figure 4 presents the fathers' mean BMI by region. The South-West region had the lowest mean consistently across the four surveys, averaging about 24.30 kgm-2.The regions also roughly fell into three groups with the highest mean BMI in Wales, West Midlands, North and Yorkshire and Humberside, a middle group of North West, East Midlands and Scotland, and the lowest BMI group of East and South-East and South West.

1.2. Regional Migration

Fathers were classified as intra-generational non-migrants if they remained in the same region between 1958 and 1974, and migrants if they moved to another region over this time period. The same classification was used for sons between 1991-2000. Inter-generational father-son regional migration was also examined between 1958-2000.

Table 2 presents the results of the analyses for height, weight and BMI respectively between geographically mobile and non-migrant groups. Intra-generational regional migration was 9.3% for fathers and 6.7% for sons, while inter-generational regional migration was 22.2%. There was a general trend for intra-generational migrants to be taller than non-migrants, by on average from 1.4cm in fathers to 2cm in sons. The inter-generational difference was 0.66cm and was borderline significant. There were no significant differences in mean weights between migrants and non-migrants in either geographically mobile groups both intra- and inter-generationally. For BMI migrating fathers had, on average, a lower BMI than sedentes but there was no significant intra-generational difference for sons although the directionality was maintained. Analysis of the inter-generational migration revealed that migrants had, on average, a lower BMI.

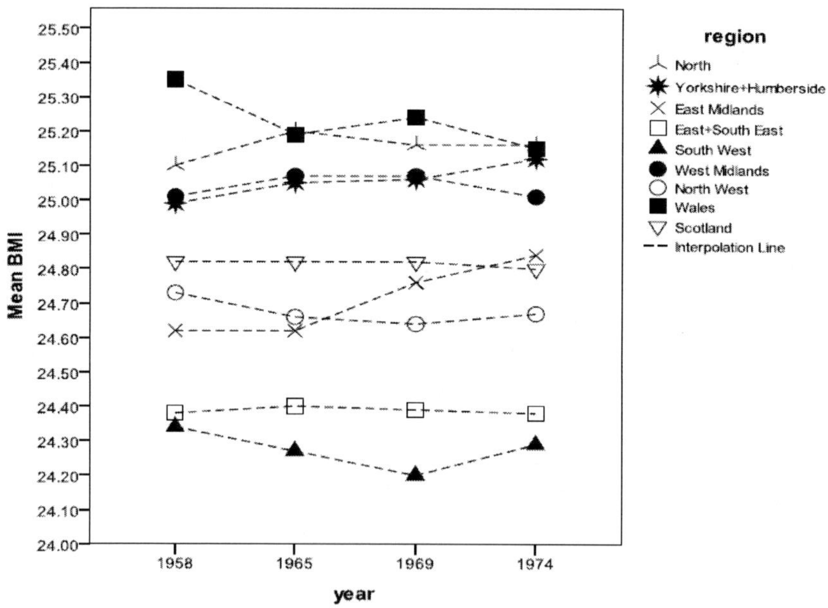

Figure 4. Fathers' mean BMI by region (Krzyżanowska and Mascie-Taylor, 2011b).

The pattern of migration was examined further by distinguishing between in- and out-migrants. The mean differences in height, weight and BMI between in-migrants and non-migrants and out-migrants and non-migrants in height, weight and BMI were analysed and the net effect of migration (the difference between in- and out-migrants) was also calculated.

Table 2. Mean height, weight and BMI for non-migrants and migrants (Krzyżanowska and Mascie-Taylor, 2011b)

Geographical migration	Height					Weight				BMI			
Fathers:1958-1974	N	Mean	%	*t*	*p*	N	Mean	*t*	*p*	N	Mean	*t*	*p*
Nonmigrants	1895	174.82	90.7	2.537	0.011	1888	75.62	0.273	0.785	1888	24.75	2.100	0.036
Migrants	195	176.17	9.3			194	75.41			194	24.28		
Sons:1991-2000													
Nonmigrants	1949	177.11	93.2	3.414	0.001	1919	80.07	0.361	0.718	1919	25.51	1.485	0.138
Migrants	141	179.11	6.8			138	80.47			138	25.03		
Fathers-sons:1958-2000													
Nonmigrants	1626	177.10	77.8	1.884	0.060	1600	80.32	1.549	0.122	1600	25.59	2.650	0.008
Migrants	464	177.76	22.2			457	79.29			457	25.08		

Figure 5. Mean difference in fathers' height among in- (A) and out-migrants (B) and the net effect (C) (Y+H=Yorkshire+Humberside; East M=East Midlands; West M=West Midlands; NW= North West) (Krzyżanowska and Mascie-Taylor, 2011b).

Figures 5 to 8 show these results just for fathers' and sons' height and BMI as there were no significant differences for fathers' and sons' weight and so no figures are shown. A positive sign indicates that migrants were taller or had a higher BMI, on average, than non-migrants or that the net effect of in- and out-migration was positive. For fathers, in-migrants for height were taller, on average in 6 of the 9 regions with the greatest difference in the North and West Midlands regions. The tallest in-migrants among fathers came into the North region (3.0cm above non-migrants, Figure 5A) and the tallest out-migrants among fathers came from Yorkshire and Humberside (3.5cm taller, on average, than non-migrants) and the shortest from Scotland (0.9cm below non-migrants) (Figure 5B). The net effect of migration (Figure 5C) revealed that the largest positive gain on fathers' height was in the West Midlands region and Scotland (3.38cm and 3.28 cm, respectively) while negative effects were found in Yorkshire and Humberside, East Midlands, East and South East by between 0.75cm and 3.72cm.

For sons (Figure 6A) in-migrants were, on average, taller in all regions except Wales, with the largest differences between in-migrants and non-migrants in East and the South East (+3.69cm) and South-West (+2.56).For out-migrants (Figure 6B), the tallest (3.36cm above sedentes) sons came from Wales, while the shortest (1.3cm below non-migrants) from the East Midlands region. The net effect of migration showed that the North, East Midlands, East and South East and West Midlands regions were net gainers (Figure 6C) while Wales and Scotland, in particular, were net losers.

For BMI among fathers, in-migrants (Figure 7A) in 7 of the 9 regions were, on average, of lower BMI than non-migrants, particularly in Wales and Scotland (-2.14 and -1.82 kgm^{-2}, respectively). For out-migrants (Figure 7B) fathers from the North West (+0.10 above the mean for sedentes) and South West regions (+0.33 above) were the only two regions showing positive values and the largest negative values were found in the East Midlands and Yorkshire and Humberside (-0.95 and -0.98, respectively).The net effect of migration (Figure 7C) indicated that in 4 regions there was a positive effect, the largest gains being in the East Midlands and Yorkshire and Humberside regions (+1.38 and 1.00, respectively) and largest losses in Scotland (-1.48) and Wales (-1.46).

Figure 6. Mean difference in sons' height among in- (A) and out-migrants (B) and the net effect (C) (Krzyżanowska and Mascie-Taylor, 2011b).

Figure 7. Mean difference in fathers' BMI among in- (A) and out-migrants (B) and the net effect (C) (Krzyżanowska and Mascie-Taylor, 2011b).

For sons the results of inter-generational migration for BMI are shown in Figure 8. In-migrating sons into the North West (0.22 above) and Wales (0.42 above) had higher BMI, on average, than sedentes while in-migrants into

Yorkshire and Humberside were, on average, 2.30 lower in BMI (Figure 8A).In all regions out-migrants had lower BMI than non-migrants from -0.10 in Wales to -0.86 in Yorkshire and Humberside (Figure 8B). The net effect of migration revealed that 6 of the 9 regions were net gainers, particularly the North West (+1.04) while, Yorkshire and Humberside were net loser (-1.44, Figure 8C).

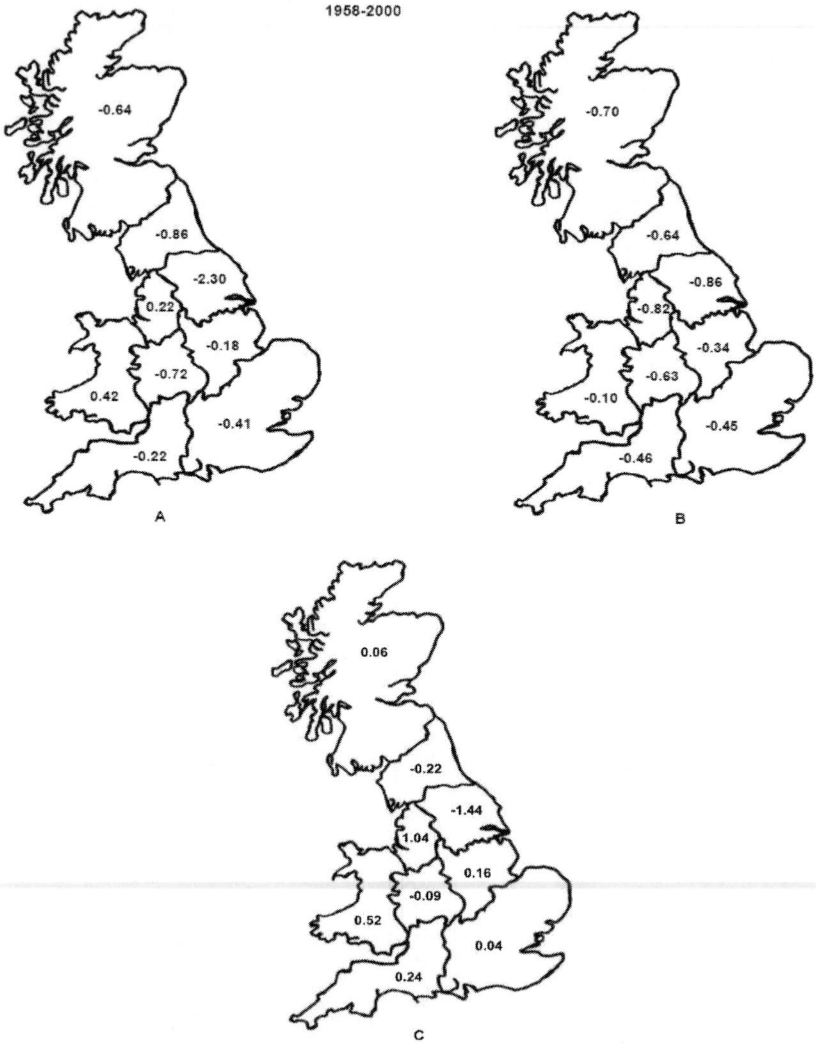

Figure 8. Mean difference in sons' BMI among in- (A) and out-migrants (B) and the net effect (C) (Krzyżanowska and Mascie-Taylor, 2011b).

1.3. Discussion of Findings

Geographical variation in anthropometric traits has been described in many countries:Canada (Reeder et al., 1997; Shields and Tjepkema, 2006), Finland (Jokela et al., 2009), India (Dey and Debray, 2003), Peru (Shin, 2007), Poland (Hulanicka et al., 1999), Sweden (Rasmussen et al., 1999), Turkey (Iseri and Arslan, 2009), United States (Jackson et al., 2005) as well as Great Britain (Rona and Altman, 1977; Rosenbaum et al., 1985; Scarborough and Allender, 2008).

The results of the current study showed highly significant regional variation in height in the fathers generation. Men from the East and South East were the tallest, while the shortest came from Wales. The overall mean height difference between regions ranged from about 2.7cm to 3.1cm (Figure 2). This geographic variation in height found in our study is consistent with earlier research (Mascie-Taylor, 1984; Mascie-Taylor and Boldsen, 1985) which showed that the shortest males were in Wales and Scotland and the tallest in the South East together with the Southern and Eastern regions. Geographical variation in height has also been reported in the British National Cohort (NCDS) by Strachan et al. (2007) as well as in other British data (Knight and Eldridge, 1984; Elford et al., 1990). In those surveys men examined in the South were taller on average than those from Scotland. Men from the North, Midlands and Wales were of intermediate height. Our findings are also in agreement with the Rosenbaum et al. (1985) study. People living in Wales were shorter than others; those living in Scotland were the second shortest group but those living in the East of England and South East were around 1 cm taller than those living in the North.

Regional differences in BMI were investigated by Waldhör et al.(2003), Ekblom et al. (2004), Schober et al.(2007), Strachan et al. (2007). Higher mean BMI was reported in small towns or the countryside in Sweden (Ekblom et al., 2004). A clear east-west gradient in BMI was observed in Austria (Waldhör et al., 2003; Schober et al., 2007). There were some areas in the south and north-west of Spain with the highest BMI than the rest of the country (Gutiérrez-Fisac et al., 1999). However, there are some examples where an 'obesogenic environment' seems to be similar across different degrees of urbanization(Gast et al., 2007; Peytremann-Bridevaux andSantos-Egglma, 2007) or across geographical areas (Lahti-Koski et al., 2008). In our study there were much greater differences in mean BMI between regions (Figure 4) and less in weight (Figure 3).The highest mean weight was generally in the West Midlands, while the lowest was in the South West region. For BMI the highest mean value was in Wales and the North, and the lowest, similar to weight, in the South West region. The highest BMI in Wales and Scotland was found by Strachan et al. (2007). Similar results, especially for BMI, were reported by Rosenbaum et al. (1985). Rennie and Jebb (2005) observed small regional differences in overweight and obesity within England. They have noticed that the prevalence of obesity in Scotland and Wales was greater than in England.

Studies on the relationship between geographical migration and body height are well-known (Kobyliansky and Arensburg, 1977; Kim, 1982; Wheeler and Tan, 1983; Bernis, 1984; Steegmann, 1985; Singh and Harrison, 1996) and many researchers have found that both internal and international migrants are taller than non-migrants (Bernis, 1984; Mascie-Taylor, 1984; Danubio et al., 2005; Szklarska et al., 2008), which was clearly confirmed in our study. Migrants were taller, on average, than non-migrants by 1.4cm in fathers and 2cm in sons (Table 2). Similar results on NCDS data were reported by Mascie-Taylor (1984) and Strachan et al. (2007).

The relation between geographical migration and BMI was also studied but results are not consistent between studies. For example, Danubio et al. (2005) pointed out that the Italian immigrants to USA had a higher prevalence of obesity than the contemporary US sample, while Park et al. (2009) found a lower chance of obesity among all immigrants. Tunisian migrants into France were less overweight than French (Méjean et al., 2007), while migrant children of Turkish, Russian and Polish origin had higher prevalence of overweight and obesity than those native-born in Germany (Will et al., 2005). Similarly, internal Senegalese migrants had a higher BMI than non-migrants (Garnier et al., 2003). Our study did not show any differences in weight between geographically mobile groups both in fathers and sons but migrating fathers had lower BMI than sedentes. Our results among fathers are consistent with the previous study on NCDS (Strachan et al., 2007).

The results based on these father-son pairs confirm that there is selective migration in relation to height and BMI. It appears that people who were geographically mobile within Britain in the second half of the twentieth century tended to be taller and have lower BMI than non-migrants.

2. Social Class Variation and Social Class Mobility in Relation to Height, Weight and Body Mass Index

2.1.Social Class Variation

Table 3 presents the fathers' mean heights by social class on four occasions (1958, 1965, 1969 and 1974) and the sons' mean heights on two occasions (1991 and 2000) for the 2090 father-son pairs. Highly significant differences in means were found between social classes over all father and son follow-up dates, and there was a general tendency for mean height to decrease from social class I to V in both fathers and sons. The overall mean height difference between social class I and V ranged from about 4 cm to 4.8 cm in the fathers' generation but declined to between 2.7 cm and 3.3 cm in the sons' generation.

There was much less variation in mean weight between social classes (Table 4), and even less for BMI (Table 5). In fathers the highest mean weight was generally in social class II, while in sons highest means were found in classes III and IV. Body mass index only differed significantly by social class in sons in 1991 and the main cause of this heterogeneity was the lower mean in social class I.

Table 3. Fathers' and sons' mean height (cm) and percentage distribution by social class (Krzyżanowska and Mascie-Taylor, 2011a)

	I		II		III		IV		V			
	Mean height	%	Mean height	%	Mean height	%	Mean height	%	Mean height	%	*F*	*p*
Fathers												
1958	178.06	4.6	176.22	16.1	174.76	61.5	173.78	11.9	173.23	5.9	11.175	<0.001
1965	177.18	6.1	176.32	16.4	174.84	57.5	173.68	15.6	172.50	4.4	11.364	<0.001
1969	177.37	6.7	176.42	19.6	174.64	54.0	173.59	14.9	173.05	4.7	13.927	<0.001
1974	176.84	6.1	176.08	23.0	174.72	55.0	173.45	12.9	172.80	3.0	10.170	<0.001
Sons												
1991	178.59	9.8	177.81	31.8	176.75	45.6	176.78	10.8	175.28	1.9	5.682	<0.001
2000	178.55	7.4	177.79	40.8	176.69	42.2	176.54	7.4	175.85	2.2	5.307	<0.001

Table 4. Fathers' and sons' mean weight (kg) by social class (Krzyżanowska and Mascie-Taylor, 2011a)

	I	II	III	IV	V		
	Mean weight	Mean weight	Mean weight	Mean weight	Mean weight	*F*	*p*
Fathers							
1958	77.30	76.62	75.21	75.48	75.77	2.100	ns
1965	76.71	77.04	75.24	75.20	74.82	2.859	0.022
1969	75.93	77.30	75.33	74.92	73.25	5.017	0.001
1974	76.21	77.11	75.12	74.74	75.15	4.095	0.003
Sons							
1991	78.08	79.79	80.80	80.39	77.32	2.662	0.031
2000	78.04	80.09	80.42	80.90	78.28	1.562	ns

Table 5. Fathers' and sons' mean BMI (kg/m^2) by social class (Krzyżanowska and Mascie-Taylor, 2011a)

	I	II	III	IV	V		
	Mean BMI	Mean BMI	Mean BMI	Mean BMI	Mean BMI	*F*	*p*
Fathers							
1958	24.37	24.68	24.63	24.98	25.24	2.092	ns
1965	24.43	24.80	24.61	24.93	25.18	1.736	ns
1969	24.16	24.83	24.71	24.86	24.47	1.740	ns
1974	24.38	24.86	24.62	24.87	25.15	1.521	ns
Sons							
1991	24.47	25.22	25.84	25.74	25.14	7.395	<0.001
2000	25.83	26.66	26.69	26.77	26.82	1.285	ns

2.2. Social Mobility

Fathers and sons were classified as non-mobile if they remained in the same social class between time periods, downwardly mobile if they moved down one or more classes between time periods, and upwardly mobile if they moved up one or more classes over this time period. The non-mobile group were set as the reference group and assigned a value of zero. The mean height, weight and BMI values presented in Tables 6 to 8 for upward and downward mobile groups are relative to the non-mobile group.

Table 6. Extent of intra- and inter-generational mobility of fathers and sons in relation to height (Krzyżanowska and Mascie-Taylor, 2011a)

	Downward		Upward			
	Mean[a]	%	Mean[a]	%	*F*	*p*
Intra-generational						
Fathers						
1958–1965	–0.59	18.6	1.29	19.3	7.600	0.001
1958–1969	–0.45	18.1	1.67	22.6	12.164	<0.001
1965–1969	–0.70	12.6	0.88	15.9	3.755	0.024
1965–1974	–0.29	11.4	0.59	20.9	1.512	ns
1969–1974	0.88	11.3	1.61	17.7	8.319	<0.001
1958–1974	–0.10	15.4	1.62	26.2	10.586	<0.001
Sons						
1991–2000	–0.71	15.0	0.80	21.0	4.791	0.008
Inter-generational (fathers–sons)						
1965–1991	–0.94	18.2	1.20	40.6	13.302	<0.001
1974–2000	–0.90	19.4	0.59	39.0	6.295	0.002

[a]Mean height difference (cm) from non-mobile reference group (set to zero).

There was a general trend that downwardly mobile fathers were, on average, shorter than the non-mobile group by between 0.1 cm and 0.7 cm, while the upwardly mobile group were taller by between 0.6 cm and 1.7 cm. The only exception was the social mobility between 1969 and 1974, in which both downwardly and upwardly mobile groups were taller than the non-mobile group (respectively by between 0.9 cm and 1.6 cm). Over the whole intra-generational time period from 1958 to 1974, the main finding was of upwardly mobile fathers being taller, on average, than non-mobile fathers, and for downwardly mobile fathers to be slightly shorter. For sons, downwardly mobile sons were significantly shorter, while upwardly mobile were taller than non-mobile sons (Table 6).

Table 7. Extent of intra- and inter-generational mobility of fathers and sons in relation to weight (Krzyżanowska and Mascie-Taylor, 2011a)

	Downward	Upward		
	Mean[a]	Mean[a]	*F*	*p*
Intra-generational				
Fathers				
1958–1965	–0.92	0.62	2.565	ns
1958–1969	–1.36	0.44	3.812	0.022

Table 7. (Continued)

	Downward	Upward		
	Mean[a]	Mean[a]	*F*	*p*
1965–1969	–0.65	0.68	1.280	ns
1965–1974	–0.54	0.26	0.498	ns
1969–1974	–0.31	0.10	0.138	ns
1958–1974	–0.86	0.41	1.674	ns
Sons				
1991–2000	0.93	1.41	2.315	ns
Inter-generational (fathers–sons)				
1965–1991	0.24	0.70	0.640	ns
1974–2000	1.18	–0.18	1.424	ns

[a]Mean weight difference (kg) from non-mobile reference group (set to zero).

There were no significant intra-generational differences in weight between socially mobile groups except the mobility of fathers between 1958 and 1969, where the downwardly mobile were lighter, on average, by 1.4 kg, while the upwardly mobile group were heavier by 0.4 kg. However, there was a tendency for downwardly mobile fathers to be lighter than the non-mobile group, while the upwardly mobile were heavier (Table 7).

The results of intra-generational mobility for BMI are shown in Table 8. Most of the analyses (with the exception of the 1958–1965 data) were non-significant and both upward and downward mobile groups had a lower mean BMI than the non-mobile.

Table 8. Extent of intra- and inter-generational mobility of fathers and sons in relation to BMI (Krzyżanowska and Mascie-Taylor, 2011a)

	Downward	Upward		
	Mean[a]	Mean[a]	*F*	*p*
Intra-generational				
Fathers				
1958–1965	–0.11	–0.11	0.362	ns
1958–1969	–0.34	–0.33	3.294	0.037
1965–1969	–0.02	0.00	0.005	ns
1965–1974	–0.07	–0.09	0.182	ns
1969–1974	–0.31	–0.43	4.032	0.018
1958–1974	–0.23	–0.31	2.514	ns

	Downward	Upward		
	Mean[a]	Mean[a]	*F*	*p*
Sons				
1991–2000	0.50	0.23	2.572	ns
Inter-generational (fathers–sons)				
1965–1991	0.30	–0.12	1.484	ns
1974–2000	0.68	–0.07	6.254	0.002

[a]Mean BMI difference (kg/m^2) from non-mobile reference group (set to zero).

Inter-generational mobility was highly significant for height with downwardly mobile sons having, on average, a lower mean stature by about 0.9 cm, while upwardly mobile sons were taller, on average, by between 0.6 cm and 1.2 cm. There were no significant inter-generational weight effects, while for BMI downwardly mobile sons had a higher BMI, on average, while upwardly mobile sons had a slightly lower BMI, on average, than non-mobile sons.

2.3. Discussion of Findings

Strong social inequalities in height in a British generation born in 1958 have been well documented (Lasker and Mascie-Taylor, 1989; Terrell and Mascie-Taylor, 1991; Power et al., 2002; Li et al., 2004; Li and Power, 2004). This study shows that differences in mean height between the extreme social classes (I and V) were highly significant but greater for fathers than for sons, and that the differences decreased over time within each generation. For example, for fathers the height difference between social classes I and V was 4.8 cm in 1958 and 4.0 cm in 1974; for sons, it was 3.3 cm in 1991 and 2.7 cm in 2000 (Table 3). The reduction in the extent of social class mean differences in height has also been noted by Li et al. (2004) and Li and Power (2004).

Studies on the relationship between social mobility and body height are well known in the literature. Many researchers have found that tall people and those with better health are more mobile, more often socially promoted and tend to occupy higher positions that require better qualifications (Mascie-Taylor, 1984; Lasker and Mascie-Taylor, 1989; Cernerud, 1995; Bielicki and Szklarska, 2000; Power et al., 2002).

These results on father–son pairs suggest that there is some selective social mobility in relation to height. It appears that people who are upwardly mobile (who have moved up the social scale) tend to be taller than non-mobile ones, while downwardly mobile tend to be shorter.

The inter-generational mobility for BMI suggests a different scenario with the downwardly mobile having a higher BMI, and a risk of obesity in the lower social classes. These findings are in agreement with Swedish data, which also show that obese men are often downwardly mobile (Karnehed et al., 2008).

The general conclusion from this study is that there is a trend of diminishing inequalities in height between social classes, but that BMI differences between social classes may be increasing.

3. Biosocial Correlates of Intergenerational Social Mobility

3.1. Introduction

Both sons and daughters were studied. Birth order was categorised as first, second and third born and above. Father's social class was defined as before as was the extent of inter –generational mobility i.e., none (no change in social class between the father's and index child's social class), upwardly mobile (where the index child moved up one or more social classes compared with their father) or downwardly mobile (where the index child moved down one or more social classes compared with their father).Because sons born in social class I can only be non-mobile or downwardly mobile and those sons born in social class V can only be non-mobile or upwardly mobile, the analyses were restricted to fathers social classes II to IV. Families were placed into one of the 9 standard regions. Education of the index child as well as father's educational level were classified as: 1, university; 2,'A' level(s); 3,'O' level; 4, apprenticeship or no qualifications. The reading, mathematics, verbal and non-verbal IQ test scores of the index child at age 11 were also analysed. Family size was analysed as a continuous variable and because of right skew was normalized using a logarithmic (base 10) transformation.

3.2. Relationship between Biosocial Variables and Inter-Generational Social Mobility

Table 9 presents the extent of inter-generational social mobility of sons and daughters in relation to five categorical biosocial variables. Birth order and social mobility were just significantly associated but only for sons ($\chi^2_{(4)}$=11.28, p=0.024). The main finding was that first born sons were more likely to be socially non-mobile or upwardly mobile than expected while second or later born were more likely to be downwardly socially mobile (Table 9).

There was a significant relationship between father's social class and inter-generational social mobility of sons ($\chi^2_{(8)}$=421.31, p<0.001) and daughters ($\chi^2_{(8)}$=364.02, p<0.001) (Table 9). Children from social class III and IV were more likely to be upwardly socially mobile than expected and those from social class II were much more likely to be downwardly socially mobile than expected (Table 9).

The results for education of the cohort members (whether sons or daughters) indicated that there was a very significant association between their educational attainment and inter-generational social mobility ($\chi^2_{(6)}$=246.14, p<0.001 for sons and $\chi^2_{(6)}$=311.04, p<0.001 for daughters). It appeared that those with higher educational qualifications (categories 1 and 2) were much more upwardly socially mobile than expected while children with medium level of education and without qualifications (categories 3 and 4) were more likely to be non-socially mobile or downwardly mobile than expected (Table 9).

There were significant relationships between father's education and inter-generational social mobility of sons ($\chi^2_{(6)}$=14.52, p=0.024). Those whose fathers attained the lowest level of education were more likely to be non-socially mobile (29.0%) or upwardly mobile (23.2%) than downwardly mobile. Furthermore, sons whose fathers achieved higher educational level were much more downwardly mobile than expected (Table 9).

No regional differences were found in inter-generational social mobility ($\chi^2_{(16)}$=15.59, p=0.482 for sons and $\chi^2_{(16)}$=12.65, p=0.698 for daughters) (Table 9).

There was a significant relationship between the number of children in the family and inter-generational social mobility of the cohort members (F=4.69, p=0.009 for sons and F=5.14, p=0.006 for daughters) (Table 10). Upwardly socially mobile or non-mobile children came from families with smaller numbers of children, while children with higher number of siblings were much more likely to be downwardly mobile. The analyses of the cognitive tests revealed strong associations with social mobility, more so for sons than daughters (Table 10). There was a general tendency for mean scores to increase from downwardly socially mobile to upwardly mobile. Post hoc tests revealed that the main differences were between the upwardly mobile versus the non and downwardly mobile, for example, reading score the upwardly mobile mean was 2.31 and 3.30 above the non and downwardly mobile, whereas the difference between the non-mobile and downwardly mobile was only 0.99.In order to test which of the biosocial variables contributed to explaining inter-generational social mobility, a series of multinomial logistic regression analyses were undertaken for sons and daughters separately. In the first analysis all ten socioeconomic and cognitive variables were included in the model. It appeared that in both sons and daughters their education and social class of the father as well as non-verbal IQ, reading and mathematics score for sons were the most important variables. Overall just over 60% of the three mobility categories were predicted for both sons and daughters, and there was better prediction of the non-mobile, followed by the upwardly and downwardly mobile groups.

In the second analysis a stepwise multinomial logistic regression procedure was used in which the most significant predictor entered the model first followed by the second most significant predictor etc. The results (Table 11) revealed that in both sons and daughters the most significant predictor of mobility categories was education of the cohort member, followed by social class of the father and then by mathematics score. Only in sons were reading and non-verbal IQ scores significant predictors. With this reduced model the percentage correctly predicted fell slightly in both sons and daughters with the non-mobile group being best predicted. Better educated cohort members were (1.4 to 8.3 for sons and 1.4 to 18.4 for daughters) times more likely to be upwardly mobile than not qualified. Sons and daughters of fathers from higher social classes were 4.0 to 23.1 for sons and 3.8 to 20.8 for daughters times more likely to be downwardly mobile than children of fathers from lower social class (Table 11).

Table 9. Relationship between categorical variables and inter-generational social mobility (row percentages are presented) (Krzyżanowska and Mascie-Taylor, 2013)

Variable	Sons					Daughters				
	Social mobility (%)			$\chi2$	p	Social mobility (%)			$\chi2$	p
	Down	Non	Up			Down	Non	Up		
Birth order										
1	13.3	45.0	41.7	11.28	0.024	25.3	40.8	33.9	2.20	ns
2	17.6	44.7	37.7			24.1	42.7	33.2		
3+	18.5	45.9	35.6			27.1	41.2	31.7		
Father's social class										
II	39.4	47.3	13.3	421.31	<0.001	52.6	38.8	8.6	364.02	<0.001
III	13.6	50.1	36.3			23.3	44.9	31.8		
IV	1.7	16.4	81.9			5.0	27.7	67.3		
Education of CM*										
University	7.2	26.7	66.1	246.14	<0.001	10.0	25.6	64.4	311.04	<0.001
'A' level(s)	13.7	47.1	39.2			22.0	40.2	37.8		
'O' level	14.5	53.2	32.3			25.0	49.0	26.0		
No qualifications	29.1	48.5	22.4			38.0	42.5	19.5		
Education of father										
University	20.2	49.5	30.3	14.52	0.024	28.3	37.9	33.8	5.75	ns
'A' level(s)	17.4	37.8	44.8			29.3	37.5	33.2		
'O' level	13.6	45.0	41.4			25.7	41.6	32.7		
No qualifications	16.4	46.5	37.1			24.2	42.8	33.0		

Table 9. (Continued)

Variable	Sons					Daughters				
	Social mobility (%)			$\chi 2$	p	Social mobility (%)			$\chi 2$	p
	Down	Non	Up			Down	Non	Up		
Region										
North	17.8	46.1	36.1	15.59	ns	28.2	44.8	27.0	12.65	ns
Y + H**	17.1	44.5	38.4			25.3	41.6	33.1		
East Midlands	11.0	48.8	40.2			29.8	38.0	32.2		
East+South East	16.9	44.3	38.8			24.5	42.6	32.9		
South West	16.9	48.1	35.0			28.6	38.8	32.6		
West Midlands	14.9	50.5	34.6			26.4	41.7	31.9		
North West	19.0	42.1	38.9			25.0	41.8	33.2		
Wales	12.0	48.9	39.1			25.9	42.6	31.5		
Scotland	16.9	38.5	44.6			21.1	39.1	39.8		

*CM= cohort member (son or daughter)

**Y+H=Yorkshire + Humberside

Table 10. Relationship between continuous variables and inter-generational social mobility (Krzyżanowska and Mascie-Taylor, 2013)

Variable	Sons								Daughters							
	Social mobility						*F*	*p*	Social mobility						*F*	*p*
	Down		Non		Up				Down		Non		Up			
	Mean	SD	Mean	SD	Mean	SD			Mean	SD	Mean	SD	Mean	SD		
Number of children	3.08	1.23	2.99	1.19	2.87	1.13	4.69	0.009	3.05	1.24	2.90	1.17	2.85	1.15	5.14	0.006
Verbal IQ	19.92	8.94	21.73	8.73	24.87	8.51	50.44	<0.001	22.98	8.85	24.53	7.95	26.75	7.68	37.22	<0.001
Non-verbal IQ	19.62	7.13	21.28	6.89	23.89	6.81	57.58	<0.001	21.14	7.19	22.11	6.74	23.78	6.50	26.69	<0.001
Reading score	15.24	5.92	16.23	5.74	18.54	5.86	53.93	<0.001	15.77	5.69	16.70	5.30	18.26	5.33	37.24	<0.001
Mathematics score	14.96	9.82	17.29	9.60	21.20	9.71	63.65	<0.001	16.10	9.74	17.62	9.28	20.81	9.56	44.97	<0.001

Table 11. Results of multinomial regression analyses to predict inter-generational social mobility (Krzyżanowska and Mascie-Taylor, 2013)

Reduced model		Sons				Daughters			
		χ^2	*p*	OR		χ^2	*p*	OR	
				Down	Up			Down	Up
Steps:1	**Education of CM***								
	1	457.683	<0.001	0.260	8.331	666.844	<0.001	0.148	18.347
	2			0.409	1.989			0.414	2.748
	3			0.430	1.422			0.492	1.354
	4			1	1			1	1
2	**Father's sc****								
	II	414.513	<0.001	23.063	0.009	368.720	<0.001	20.800	0.008
	III			3.996	0.068			3.760	0.142
	IV			1	1			1	1
3	**Mathematics score**	77.985	<0.001	-	-	26.990	<0.001	-	-
4	**Reading score**	19.903	<0.001	-	-	-	-	-	-
5	**Non-verbal IQ**	12.721	0.002	-	-	-	-	-	-

***CM** – cohort member (son or daughter)

****sc** – social class

(63.3% of boys and 62.2% of girls were correctly classified (boys: downwardly mobile 17.9%; non-mobile 80.9%; upwardly mobile 61.9%; girls: 34.9%; 77.4%; 64.0%, respectively).

3.3. Discussion of Findings

Among the biosocial variables analysed separately almost all of them birth order, sibship size in the families, father's social class and his education, educational attainment of the cohort members and the results of their cognitive tests (reading, mathematics, verbal and non-verbal IQ tests) associated with cohort members social class.

Birth order has been considered one of the most influential environmental factors in child development (Berger and Felsenthal-Berger, 2009). Firstborns tend to be larger, stronger, have better health outcomes and socio-economic environment (Barreto and Rodrigues, 1992; Elliott, 1992; Kaplan et al., 1992; Li and Taylor, 1993; Lewis and Britton, 1998; Modin, 2002; Argys et al., 2006), than later born siblings. A number of studies have found that firstborns had an academic advantage (Zajonc and Markus, 1975; Zajonc, 1976; Cherian, 1990; Sulloway, 1996; Kantarevic and Mechoulan, 2005; Fergusson et al., 2006; de Haan, 2010) and IQ scores (Paulhuset al., 1999; Bjerkedal et al., 2007; Boomsma et al., 2008; Black et al., 2011) over laterborns, but other studies have failed to confirm this finding (Steelman and Powell, 1985; Ha and Tam, 2011). A significant positive relationship between birth order and social skills as: outgoingness, getting along with others, popularity, ease in making friends has also been found (Steelman and Powell, 1985).

According to Sulloway (1996) children have an innate tendency to develop the attitudes and personalities best-suited for maximizing the parental resources that they get. They compete for parental investment (time, energy and material resources) by creating distinctive niches. It was hypothesized (Sulloway, 1996) that firstborns are less agreeable as compared to laterborn children because firstborns occupy a dominant position in a sibling group and develop more conservative attitudes (dependent, serious, more sensitive to parents' expectations) than their younger siblings. Firstborns have been found to be more extraverted, self-confident, assertive and aggressive than laterborns (Sulloway, 1996; Paulhuset al., 1999; Rohde et al., 2003). They are subjected to more accelerated role playing and training, more responsible, well-organized, supportive of authority, 'tough-minded' and patient in real-life decisions. Consequently, early-born children tend to be more ambitious, achievement- and success-oriented as well as motivated to attain high social position and eminence than their laterborn counterparts (Sulloway, 1996; Rohde et al., 2003; Healey and Ellis, 2007).

The findings from this study revealed that first born sons were more likely to be socially upwardly mobile or non-mobile, while later born were more likely to be downwardly socially mobile.However, in the multivariate analyses the birth order effect was not significant.

An inverse and consistent relationship between the number of siblings and health as well as educational outcomes has been widely reported. For example it was found that people who have more siblings, on average, have grown up in more overcrowded accommodation, with greater exposures to early infections, and with access to a less adequate diet (Blane et al., 1999). Consequently, greater number of siblings is strongly related to early childhood morbidity and mortality risk (Hart and Davey Smith, 2003). Furthermore, smaller sibships are more likely to produce higher achieving and more intelligent children than larger families (Blake, 1989; Falbo and Polit, 1986; Downey, 1995; Iacovou, 2001, Bradley and Taylor, 2004; Parr, 2006; Li et al., 2008; Booth and Kee, 2009; Jaeger, 2009; Sen and Clemente, 2010).

Children who come from smaller families have a better chance to be upwardly mobile compared with children from larger families. A negative relation between upward mobility and family size was found by Tomasson (1966) in a sample of 583 couples from Peoria, Springfield, and Rockford, Illinois. Marjoribanks (1987), using longitudinal data from 310 Australians, found that sibsize was not related to social mobility of males, but did influence the eventual social status of females. Decreased chance of upward mobility with increasing number of siblings has been reported by Blane et al. (1999) in a cohort of 5,645 employed men from the West of Scotland Collaborative study. Van Bavel et al. (2011) investigated the association between family size and intergenerational mobility in the city of Antwerp between 1846 and 1920 and found, after controlling for confounding factors, that people with many children were more likely to end up in the lower classes.

In the current study children from families with smaller numbers of sibs were more likely to be socially non-mobile or upwardly mobile. Although, similarly to birth order, the sibling size effect became non-significant in the multivariate analyses.

The role of social origin (parental social class in childhood) in health outcomes, educational attainment and social class in adulthood has been reported very often (Shavit and Blossfeld, 1993; Davey Smith et al., 1998; Bosma et al., 1999; Breen and Jonsson, 2000; Breen and Goldthorpe, 2001; Jaeger and Holm, 2003; Iannelli and Paterson, 2005; von Stumm et al., 2009). However, there is a debate among social scientists as to whether an individual's economic success and life opportunities more generally are determined by their own ability and effort (Saunders, 1997; 2002) or by their parents' socio-economic status (Breen and Jonsson, 2000; Breen and

Goldthorpe, 1999; 2001). Breen and Goldthorpe (1999; 2001) using the NCDS data found both a substantial effect of class of origin on social class at age 33, and that education substantially mediated these effects. They have pointed out that individuals from lower class backgrounds have to display more merit than do individuals from higher class backgrounds to reach any given higher class level. Carneiro's et al. (2007) results based on the same material showed that children from both professional and non-manual family backgrounds exhibit significantly greater cognitive and non-cognitive skills than children from manual backgrounds. Jaeger and Holm (2003) using data from the Danish Youth Longitudinal Panel Survey found father's social class to be the strongest predictor of educational attainment followed by father's level of education and finally cognitive ability. It has been demonstrated on the data from the 2001 Scottish Household Survey, that education plays an intermediary role between class of origin and class of destination, i.e., there is still a strong direct effect of social class of origin on people's class of destination - even after taking educational attainment into account (Iannelli and Paterson, 2005). Deary et al. (2005) suggested that the correlation between childhood IQ and own social class at midlife is stronger than the correlation with father's social class. Furthermore, the association between parental social class and class of destination is weaker for more highly educated people than for less educated people. A study using conscription data and follow-ups on 36,156 Swedish men born between 1949–51 found that when adjusting for attained social position, people with a high social position of origin tend to have higher intelligence and level of education than people with a lower social position of origin (Sorjonen et al., 2011).

The current study also found a significant relationship between father's social class and inter-generational social mobility of the cohort members. Children from social class III and IV were more likely to be upwardly socially mobile than expected and those from social class II were much more likely to be downwardly socially mobile. In the multivariate analysis it appeared that for both sons and daughters the most significant predictor of mobility categories was education of children, followed by social class of the father.

Education has been described as 'a vehicle for professionalisation' and associated with social mobility (Müller and Shavit, 1998; Nettle, 2003; Cardano et al., 2004; Machin 2004; Deary et al., 2005; Iannelli and Paterson, 2005), measured either as number of years of education, as level of education or as IQ. In educational attainment studies, parents' educational characteristics have not been a central concern, often being viewed as one aspect of the concept of 'family background' (Sieben and De Graaf, 2001). Parental education correlates with several other factors such as income, occupation, culture and social resources that matter in a life-cycle perspective. Lampard's (2007) examination of the direct and indirect effects of parents' educations and classes on children's occupational attainment in a British Household Panel Survey (BHPS) data has shown that parental education plays a substantial role in the intergenerational transmission of advantage, and has indicated that part (but not all) of the relationship between class origin and occupational attainment can be explained in terms of the intergenerational transmission of cultural capital. Data from the Columbia County Longitudinal Study has revealed a strong relationship between parents' educational level and children occupation. Parents' educational level when the child was 8 years old significantly predicted educational and occupational success for the child 40 years later (Dubow et al., 2009).

Black et al. (2007) pointed out that IQ is a strong predictor of educational attainment and future labor market success. Studies of intelligence and class mobility have demonstrated that men who are upwardly mobile in socio-economic terms tend to have higher IQ scores than their fathers, whilst those who are downwardly mobile tend to have lower scores (Mascie-Taylor and Gibson, 1978; Waller, 1971). In the current study, higher IQ (whether sons or daughters) was associated with upward mobility in the univariate analyses and was significant for mathematics score among sons and daughters in the multivariate analyses. Reading score and non-verbal IQ were significant only among sons. Using the same NCDS data, Saunders (2002) found that class of origin accounted for about 25% of the explained variance in class at age 33, whereas ability, motivation and qualifications accounted for over 60%.

As has been reported by Nettle (2003), attained social class in adulthood was more strongly related to GA score (the General Ability test) than parental social class. Similar results were reported by Deary et al. (2005) among men who participated in the Scottish Mental Survey of 1932 and thereafter in the Midspan Collaborative study in Scotland between 1970 and 1973. It appeared that height at mid-life, years of education, and childhood IQ were significantly positively related to upward social mobility. Controlling for the effects of the individual independent variables reduced the effects, with only height and education remaining significant - childhood IQ was of borderline significance. In a birth cohort of 6281 men from Aberdeen, Scotland von Stumm et al. (2010) confirmed that education had the strongest direct effect on status attainment at midlife. The results of the present study are also in agreement with other British data analysed by Forrest et al. (2011). For the Newcastle Thousand Families 1947 birth cohort childhood IQ and achieved education level were both significantly and positively related to upward inter-generational mobility at the univariate level. However, only childhood IQ was significantly associated with upward mobility between 25 and 49-51 years, whereas only education level was significantly associated with upward mobility between 5 and 25 years, which means that IQ level is more strongly connected with social class attained in middle-age rather than early adulthood (Forrest et al., 2011).

The current results showed clearly that in both sons and daughters the most significant predictor of inter-generational social mobility was their own educational level. Children (whether sons or daughters) from social class III and IV with higher educational qualifications (having a degree or above; teaching or nursing qualification; passed 'A' level) were more likely to be upwardly socially mobile or non-mobile. In the light of these results, there is agreement with Saunders's (1997; 2002) point of view that social mobility in Britain takes place largely on meritocratic principles.

4. Impact of Both Social Mobility and Regional Migration on Variation in Male Height, Weight and Body Mass Index

4.1. Inter-Relationships between Social Mobility and Geographical Migration

Table 12 presents the extent of intra-generational social mobility and geographical migration of 2090 fathers and sons. There was much more social mobility than geographical migration, but that might partly be a consequence of using the region as the unit, as information on within-region migration was not available. Between 1958 and 1965, 38.0% of fathers were socially mobile but only 7.0% were geographical migrants. Social mobility and geographical migration were not independent in the fathers' generation ($\chi^2_{(2)}$=15.36, p<0.001), and socially non-mobile fathers were more likely also to be geographically non-migrant while socially mobile fathers (whether upwardly or downwardly mobile) were more likely to be geographical migrants than expected (Table 12).

Table 12. The relationships between intra-generational social mobility and regional migration (percentages) (Krzyżanowska and Mascie-Taylor, 2012)

		Social mobility	Geographical migration Non-migrant	Migrant	Total	χ^2	*p*
Father	**1958–65**	Down	17.0	1.6	18.6	15.36	**<0.001**
		None	58.7	3.3	62.0		
		Up	17.3	2.1	19.4		
		Total	93.0	7.0	100		
	1958–69	Down	16.4	1.7	18.1	6.80	**0.033**
		None	55.0	4.3	59.3		
		Up	20.1	2.5	22.6		
		Total	91.5	8.5	100		
	1958–74	Down	14.1	1.3	15.4	10.58	**0.005**
		None	53.7	4.6	58.3		
		Up	22.9	3.4	26.3		
		Total	90.7	9.3	100		
Son	**1991–2000**	Down	13.7	1.3	15.0	2.13	0.345
		None	59.8	4.2	64.0		
		Up	19.7	1.3	21.0		
		Total	93.2	6.8	100		

A similar pattern was found in 1958–1969: 40.7% of fathers were socially mobile and 8.5% were geographical migrants. There were more socially mobile and geographical migrants than expected. Over the total period of father's intra-generational mobility (1958–1974) 41.7% were socially mobile and 9.3% were geographical migrants. The socially non-mobile also tended to be geographical non-migrants, but unlike the previous two time periods, it was only upwardly socially mobile fathers who were more likely to be geographical migrants (Table 12). Among sons no significant interaction was found ($\chi^2_{(2)}$=2.13, p=0.345) between social and geographical mobility (1991–2000) (Table 12).

The inter-generational mobility between fathers and their sons revealed significant interactions between social and geographical mobility both for 1965–1991 and 1958–2000 ($\chi^2_{(2)}$=9.38, p=0.009, and $\chi^2_{(2)}$=7.30, p=0.026, respectively) (Table 13).

For 1965–1991, 58.9% of sons were socially mobile and 18.8% were geographical migrants, and a similar pattern was found for 1958–2000 (59.9% and 22.2%, respectively). There was an opposite relationship for downwardly mobile migrants. For 1965–1991 the socially non-mobile tended to be geographical non-migrants, upwardly socially mobile sons were more likely to be geographical migrants, while the downwardly social mobile were not more likely to be geographical migrants (Table 13). However, for 1958–2000 a different pattern emerged in that although the non-socially mobile also tended to be non-migrants, the socially mobile sons (whether upwardly or downwardly mobile) were more likely to be geographical migrants.

Table 13. The relationships between inter-generational (father–son) social mobility and regional migration (percentages) (Krzyżanowska and Mascie-Taylor, 2012)

	Social mobility	Geographical migration			χ^2	*p*
		Non-migrant	Migrant	Total		
1965–1991	Down	14.9	3.3	18.2	9.38	**0.009**
	None	34.5	6.6	41.1		
	Up	31.8	8.9	40.7		
	Total	81.2	18.8	100		
1958–2000	Down	11.9	3.7	15.6	7.30	**0.026**
	None	32.4	7.7	40.1		
	Up	33.5	10.8	44.3		
	Total	77.8	22.2	100		

4.2. Impact of Social Mobility and Geographical Migration on Variation in Height, Weight and BMI

In order to examine the inter-relationships of social mobility and geographical migration on the anthropometric variables a series of sequential multiple regression analyses were performed in which (a) the effect of social mobility was entered into the model initially followed by geographical migration and finally the interaction effect between social mobility and geographical migration, and (b) the effect of geographical migration was entered first, followed by social mobility and finally the interaction effect.

Tables 14–16 present the mean height, weight and BMI variation by social mobility and geographic migration (the non-socially mobile and non-migrants are the reference groups and have been set to a mean of zero) based on first entry into the model. A significant relationship was found between height, BMI and social mobility (Tables 14 and 16). The upwardly mobile were, on average, taller by between 1.2 and 1.6 cm and lighter by 0.21 kg m^{-2} in the fathers' generation, and about 0.7 cm in the sons' generation, while the downwardly mobile fathers were shorter by between 1.5 and 2 cm and their sons by 1.1 cm (Tables 14 and 15). Inter-generational upwardly socially mobile sons were, on average, taller by 0.4 cm for 1958–2000 and 1 cm for 1965–1991, and downwardly mobile sons were shorter by 1.6 cm and had also a higher BMI (Tables 14 and 16).

Table 14. Mean height differences by social mobility and geographic migration (Krzyżanowska and Mascie-Taylor, 2012)

		Social mobility			Geographical migration		
		Down	Up	*p*	Migrants	*p*	Interaction
Father	1958–65	−1.91	+1.24	**<0.001**	+0.85	ns	ns
	1958–69	−2.02	+1.60	**<0.001**	+0.96	ns	ns
	1958–74	−1.47	+1.35	**<0.001**	+0.54	ns	ns
Son	1991–2000	−1.14	+0.68	**0.008**	+1.59	**0.015**	ns
Father–son	1965–1991	−1.59	+1.04	**<0.001**	+0.06	ns	ns
	1958–2000	−1.57	+0.36	**0.010**	+0.32	ns	ns

Non-mobile reference group set to zero; ns, not significant.

Table 15. Mean weight differences by social mobility and geographic migration (Krzyżanowska and Mascie-Taylor, 2012)

		Social mobility			Geographical migration		
		Down	Up	*p*	Migrants	*p*	Interaction
Father	1958–65	−1.82	+0.95	ns	−1.20	ns	ns
	1958–69	−1.98	+0.63	**0.014**	−0.61	ns	ns
	1958–74	−1.92	+1.05	ns	−0.83	ns	ns
Son	1991–2000	+0.11	+1.10	ns	+0.94	ns	ns
Father–son	1965–1991	+0.48	-0.61	ns	−1.95	**0.025**	**0.015**
	1958–2000	+0.69	+0.24	ns	−0.81	ns	ns

Non-mobile reference group set to zero; ns, not significant.

Table 16. Mean BMI differences by social mobility and geographic migration (Krzyżanowska and Mascie-Taylor, 2012)

		Social mobility			Geographical migration		
		Down	Up	*p*	Migrants	*p*	Interaction
Father	1958–65	–0.08	+0.02	ns	–0.63	**0.014**	ns
	1958–69	–0.11	–0.24	**0.040**	–0.48	**0.025**	ns
	1958–74	–0.18	–0.05	ns	–0.42	ns	ns
Son	1991–2000	–0.34	+0.16	ns	–0.23	ns	ns
Father–son	1965–1991	+0.63	–0.52	**0.031**	–0.60	**0.005**	ns
	1958–2000	+0.69	–0.04	**0.002**	–0.35	ns	ns

Non-mobile reference group set to zero; ns, not significant.

There was a significant relationship between BMI and geographic migration. Migrating fathers had a lower BMI than sedentes, and this held for their sons migrating between 1965 and 1991 (Table 16).

In general, there was no significant relationship between weight and social mobility and geographic migration except for intra-generational mobility in 1958–1969 and inter-generational migration in 1965–1991 (Table 15). Upwardly mobile fathers were heavier than non-mobile ones by, on average, 0.6 kg, and the downwardly mobile were lighter by 2 kg. Migrating sons were also lighter than non-mobile ones by, on average, 2 kg (Table 15).

In the statistical analyses there was no significant interaction, which indicated that social mobility and geographical migration were acting in a simple additive way on height, weight and BMI.

4.3. Discussion of Findings

It is well known that spatial mobility is highly socially specific, with higher social groups being more mobile (Savage, 1988; Fielding, 1992; Ringdal, 1993). In keeping with these findings and a previous study on the NCDS (Mascie-Taylor, 1984), these results clearly show that social mobility and geographical migration are not independent, with socially non-mobile fathers and sons being more likely to stay in the same region, and fathers and sons who are socially mobile being more likely to move to another region. What is more, there was much more social mobility than geographical migration, but that might partly be because the geographical unit used was a standard region, and no information on intra-regional movement was available.

Height and BMI have been found to be significantly associated with social mobility and geographic migration when both kinds of mobility are considered separately (Krzyżanowska and Mascie-Taylor, 2011a, b). People who were upwardly mobile were found to be taller than non-mobile ones, while the downwardly mobile were, on average, shorter and heavier (Krzyżanowska and Mascie-Taylor, 2011a). Geographical migrants were, on average, taller than non-migrants for both fathers and sons, but BMI was only significant in migrant fathers who had, on average, a lower BMI. No significant differences were found in weight between geographically or socially mobile groups (Krzyżanowska and Mascie-Taylor, 2011a, b).

When the dual effect of social and geographical mobility on anthropometric traits was examined it appeared that height and BMI were significantly associated with social mobility and to some extent with geographical migration. The upwardly socially mobile were, on average, taller and had lower BMI values than the non-mobile, and geographical migrants had lower BMI values than non-migrants. Weight did not show any relationship with geographic and social mobility, except for fathers' mobility between 1958 and 1969 and sons' migration between 1965-1991, which is in agreement with previous studies (Krzyżanowska and Mascie-Taylor, 2011a, b).

The effects of geographical and social mobility on height and weight were analysed in the British population by Mascie-Taylor (1984). He found that there were no significant interactions between social mobility and geographical migration in relation to height and weight, and the present results are consistent with those findings as well as extending the notion of additivity to BMI.

To summarize, it is seen, particularly for height and BMI, that selective migration is more pronounced for social mobility than for geographical migration. Such selectivity contributes to the maintenance of social and geographical variation in height and BMI in the British population.

References

Argys, Laura M., Rees, Daniel I., Averett, Susan L., and Benjama Witoonchart.2006. "Birth order and risky adolescent behavior." *Economic Inquiry* 44:215-233.

Barreto, Thiago V., and Lages C. Rodrigues. 1992. "Factors influencing childhood immunization in an urban area of Brazil." *Journal of Epidemiology and Community Health* 46:357–361.

Batty, David G., Shipley, Martin J., Gunnell, David, Huxley, Rachel, Kivimaki, Mika, Woodward, Mark, Lee, Crystal M.Y., and George D. Smith. 2009. "Height, wealth, and health: An overview with new data from three longitudinal studies." *Economics and Human Biology* 7:137-152.

Berger, Itai, and Noorit Felsenthal–Berger. 2009. "Attention – deficit hyperactivity disorder (ADHD) and birth order." *Journal of Child Neurology* 24:692-696.

Bernis, Cristina. 1984. "Intergenerational changes in age at menarche and at menopause and in life span of contemporary man, and their biological meaning." *Studies in Human Ecology* 6:63–72.

Bielicki, Tadeusz, and Alicja Szklarska. 2000. "Are social-class differences in stature partly genetic? A hypothesis revisited." *American Journal of Human Biology* 12:97-101.

Bjerkedal, Tor, Kristensen, Petter, Skjeret Geir A., and John I. Brevik. 2007. "Intelligence test scores and birth order among young Norwegian men (conscripts) analysed within and between families." *Intelligence* 35:503–514.

Black, Sandra E., Devereux, Paul J., and Kjell G. Salvanes. 2007. "Small family, smart family? Family size and the IQ scores of young men." Working Paper no. 13336, 45. National Bureau of Economic Research, Inc:1-38.

Black, Sandra E., Devereux, Paul J., and Kjell G. Salvanes. 2011. "Older and wiser? Birth order and IQ of young men." *CESifo Economic Studies* 57:103-120.

Blake, Judith. 1989. *Family size and achievement.* Berkeley: University of California Press.

Blane, David, Smith, George D., and Carole Hart. 1999. "Some social and physical correlates ofintergenerational social mobility: Evidence from the West of Scotland Collaborative Study." *Sociology* 33:169-183.

Boomsma, Dorret I., van Beijsterveld, Toos C.E.M., Beem, Leo A., Hoekstra, Rosa A., Polderman, Tinca J.C.,and Meike Bartels. 2008. "Intelligence and birth order in boys and girls." *Intelligence* 36:630- 634.

Booth, Alison L., and Hiau J. Kee. 2009. "Birth order matters: the effect of family size and birth order on educational attainment." *Journal of Population Economics* 22:367-397.

Bosma, Hans, van de Mheen Dike H., and Johan P. Mackenbach. 1999. "Social class in childhood and general health in adulthood: questionnaire study of contribution of psychological attributes." *British Medical Journal* 318:18–22.

Bradley, Steve, and Jim Taylor. 2004. "Ethnicity, educational attainment and the transition from school." *The Manchester School* 72:317-346.

Breen, Richard, and John H. Goldthorpe. 1999. "Class inequality and meritocracy: A critique of Saunders and an alternative analysis." *British Journal of Sociology* 50:1–27.

Breen, Richard, and John H. Goldthorpe. 2001. "Class, mobility and merit: The experience of two British birth cohorts." *European Sociological Review* 17:81–101.

Breen, Richard, and Jan O. Jonsson. 2000. "Analysing educational careers: a multinomial transition model." *American Sociological Review* 65: 754-772.

Burt, Cyril. 1961. "Intelligence and social mobility." *British Journal of Statistical Psychology* 14:3-25.

Cardano, Mario, Costa, Giuseppe, and Moreno Demaria. 2004. "Social mobility and health in the Turin longitudinal study." *Social Science and Medicine* 58:1563-1574.

Cardoso, Hugo F.V., and Madalena Caninas. 2010. "Secular trends in social class differences of height, weight and BMI of boys from two schools in Lisbon, Portugal (1910-2000)." *Economics and Human Biology* 8:111-120.

Carneiro, Pedro, Crawford, Claire, and Alissa Goodman. 2007. *The impact of early cognitive and non-cognitive skills on later outcomes.* London: Centre for the Economics of Education, LSE.

Cartwright, Reed A., Hargreaves H.J., and Eric Sunderland. 1978. "Social identity and genetic variability." *Journal of Biosocial Science* 10:23-33.

Cernerud, Lars. 1995. "Height and social mobility. A study of the height of 10 year olds in relation to socio-economic background and type of formal schooling." *Scandinavian Journal of Social Medicine* 23:28-31.

Cherian, Varghese I. 1990. "Birth order and academic achievement of children in Transkei." *Psychological Reports* 66:19-24.

Danubio, Maria E., Amicone, Elisa, and Rita Vargiu. 2005. "Height and BMI of Italian immigrants to the USA, 1908-1970." *Economics and Human Biology* 3:33-43.

Deary, Ian J., Taylor, Michelle D., Hart, Carole L., Wilson, Valerie, Smith, George D., Blane, David, and John M. Starr. 2005. "Intergenerational social mobility and mid-life status attainment: influences of childhood intelligence, childhood social factors and education." *Intelligence* 33:455-472.

de Haan, Monique. 2010. "Birth order, family size and educational attainment." *Economics of Education Review* 29:576-588.

Dey, Swapan K., and Parthasarathi Debray. 2003. "A comparative study of maximal aerobic power of school boys of east and north-east regions of India." *Indian Pediatrics* 40:105-114.

Downey, Douglas B. 1995. "When bigger is not better: family size, parental resources, and children's educational performance." *American Sociological Review* 60:746-761.

Dubow, Eric F., Boxer, Paul, and L. Rowel Huesmann. 2009. "Long-term effects of parents' education on children's educational and occupational success: mediation by family interactions, child aggression, and teenage aspirations." *Merrill-Palmer Quarterly* 55:224-249.

Ekblom, Orjan B., Oddsson, Kristjan, and Bjorn T. Ekblom. 2004. "Prevalence and regional differences in overweight in 2001 and trends in BMI distribution in Swedish children from 1987 to 2001." *Scandinavian Journal of Public Health* 32:257–263.

Elford, Jonathan E., Phillips, A., Thomson, Andrew G., and Andrew G. Shaper. 1990. "Migration and geographic variations in blood pressure in Britain." *British Medical Journal* 300:291-295.

Elliott, Barbara A. 1992. "Birth order and health: Major issues." *Social Science and Medicine* 35:443–452.

Falbo, Toni, and Denise F. Polit. 1986. "A quantitative review of the only child literature: research evidence and theory development." *Psychological Bulletin* 100:176-189.

Fergusson, David M., Horwood, John L., and Jeoseph M. Boden. 2006. "Birth order and educational achievement in adolescence and young adulthood." *Australian Journal of Education* 50:122-139.

Fielding, Anthony J. 1992. "Migration and social mobility – South East England as an escalator region." *Regional Studies* 26:1-15.

Forrest, Lynne F., Hodgson, Susan, Parker, Louise, and Mark S. Pearce. 2011. "The influence of childhood IQ and education on social mobility in the Newcastle Thousand Families birth cohort." *BMC Public Health* 11: 895.

Garnier, Denis, Simondon, Kirsten B., Hoarau, Thierry, and Eric Benefice. 2003. "Impact of the health and living conditions of migrant and non-migrant Senegalese adolescent girls on their nutritional status and growth." *Public Health Nutrition* 6:535-547.

Gast, Gerrie C.M., Frenken, F.J.M., van Leest, Lilian A.T.M., Wendel-Vos, G.C. Wanda, and Wanda J.E. Bemelmans. 2007. "Intra-national variation in trends in overweight and leisure time physical activities in The Netherlands since 1980: stratification according to sex, age and urbanisation degree." *International Journal of Obesity* 31:515–520.

Gibson, John B. 1970. "Biological aspects of a high socio-economic group. I. IQ, education and social mobility." *Journal of Biosocial Science* 3:1-16.

Gibson, John B., and C.G. Nicholas Mascie-Taylor. 1973. "Biological aspects of a high socio-economic group. IQ components and social mobility." *Journal of Biosocial Science* 5:17-30.

Gibson, John B., Harrison, Geoffrey A., Hiorns, Robert W., and H.M. Macbeth. 1983. "Social mobility and psychometric variation in a group of Oxfordshire villages." *Journal of Biosocial Science* 15:193-205.

Gutiérrez-Fisac, J.L., Rodríguez Artalejo, F., Guallar-Castillon, Pilar, Banegas Benegas, Jose R., and Juan del Rey Calero. 1999. "Determinants of geographical variations in body mass index (BMI) and obesity in Spain." *International Journal of Obesity* 23:342-347.

Ha, Tshui S., and Cai L. Tam. 2011. A study of birth order, academic performance, and personality. International Conference on Social Science and Humanity IPEDR, 5, IACSIT Press, Singapore.

Harrison, Goeffrey A., and Anthony J. Boyce. 1972. "The framework of population studies." In The Structure of Human Populations, edited by Goeffrey A. Harrison, and Anthony J. Boyce, 1-16. Oxford: Clarendon Press.

Hart, Carole L., and Davey G. Smith. 2003. "Relation between number of siblings and adult mortality and stroke risk: 25 year follow up of men in the Collaborative study." *Journal of Epidemiology and Community Health* 57:385–391.

Hart, Carole L., McConnachie, Alex, Upton, Mark, and Graham Watt. 2008. "Risk factors in the Midspan family study by social class in childhood and adulthood." *International Journal of Epidemiology* 37:604-614.

Healey, Matthew D., and Bruce J. Ellis. 2007. "Birth order, conscientiousness, and openness to experience: Tests of the family-niche model of personality using a within-family methodology." *Evolution and Human Behavior* 28:55-59.

Heineck, Guido. 2006. "Height and weight in Germany, evidence from the German Socio-Economic Panel, 2002." *Economics and Human Biology* 4:359-382.

Hulanicka, Barbara, Gronkiewicz, Lidia, and Beata Ziętkiewicz. 1999. "Stature of boys post World War II migrants." *Annals of Human Biology* 26:549-559.

Iacovou, Maria. 2001. "Family composition and children's educational outcomes." *Working Paper of Institute for Social and Economic Research*. Retrieved April 2009. Colchester: University of Essex. http://www.iser.essex.ac.uk/pubs/workpaps/pdf/2001-12.pdf.

Iannelli, Cristina, and Lindsay Paterson. 2005. *Does education promote social mobility?CES Briefing No. 35*. Edinburgh: Centre for Educational Sociology, University of Edinburgh.

Iseri, Ali, and Nurullah Arslan. 2009. "Estimated anthropometric measurements of Turkish adults and effects of age and geographical regions." *International Journal of Industrial Ergonomics* 39:860-865.

Jackson, Elizabeth J., Doescher, Mark P., Jerant, Anthony F., and Gary L. Hart. 2005. "A national study of obesity prevalence and trends by type of rural county." *Journal of Rural Health* 21:140-148.

Jaeger, Mads M. 2009. "Sibship size and educational attainment: a joint test of the confluence model and the resource dilution hypothesis." *Research in Social Stratification and Mobility* 27:1-12.

Jaeger, Mads M., and Anders Holm. 2003. "Which background factors matter more in intergenerational educational attainment: social class, cultural capital or cognitive ability? A random effects approach." *Centre for Applied Microeconometrics* 1-28.

Johnson, Wendy, Brett, Caroline E., and Ian J. Deary. 2010. "The pivotal role of education in the association between ability and social class attainment: a look across three generations." *Intelligence* 38:55-65.

Jokela, Markus, Kivimäki, Mika, Elovainio, Marko, Viikari, Jorma, Raitakari, Olli T., and Lisa Keltikangas-Järvinen. 2009. "Urban/rural differences in body weight: Evidence for social selection and causation hypotheses in Finland." *Social Science and Medicine* 68:867-875.

Kantarevic, Jasmin, and Stephane Mechoulan. 2005. "Birth order, educational attainment and earnings: an investigation using the PSID." *IZA Discussion Papers* 1789, http://hdl.handle.net/10419/33423.

Kaplan, Bernice A., Mascie-Taylor, C.G. Nicholas, and Jasper Boldsen. 1992. "Birth order and health status in a British national sample." *Journal of Biosocial Science* 24:25-34.

Karnehed, Nina EK, Finn, Rasmussen, Tomas, Hemmingsson, and Tynelius, Per. 2008. "Obesity in young adulthood is related to social mobility among Swedish men." *Obesity* 16:654-658.

Kim, Yoon S. 1982. "Growth status of Korean schoolchildren in Japan." *Annals of Human Biology* 9:453–458.

Knight, Ian B., and Jack Eldridge. 1984. *The heights and weights of adults in Great Britain: report of a survey covering adults aged 16-64*. London: HMSO.

Kobyliansky, Eugene, and Baruch Arensburg. 1977. "Changes in morphology of human populations due to migration and selection." *Annals of Human Biology* 4:57–71.

Komlos, John, and Peter Kriwy. 2003. "The biological standard of living in the two Germanies." *German Economic Review* 4:459–473.

Krzyżanowska, Monika, and C.G. Nicholas Mascie-Taylor. 2011a. "Intra and intergenerational social mobility in relation to height, weight and BMI in a British national cohort." *Journal of Biosocial Science* 43:611-618.

Krzyżanowska, Monika, and C.G. Nicholas Mascie-Taylor. 2011b. "Geographical variation and migration analysis of height, weight and BMI in a British cohort study." *Journal of Biosocial Science* 43:733-749.

Krzyżanowska, Monika, and C.G. Nicholas Mascie-Taylor. 2012. "Impact of social mobility and geographical migration on variation in male height, weight and Body Mass Index in a British cohort." *Journal of Biosocial Science* 44:221-228.

Krzyżanowska, Monika, and C.G. Nicholas Mascie-Taylor. 2013. "Biosocial correlates of intergenerational social mobility in a British cohort." *Journal of Biosocial Science* 45:481-496.

Kues, Arne B. 2010. "Taller – healthier – more equal? The biological standard of living in Switzerland in the second half of the 20th century." *Economics and Human Biology* 8:67-79.

Lahti-Koski, Marjaana, Taskinen, Oli, Similä, Minna, Männistö, Satu, Laatikainen, Tina, Knekt, Paul, and Lisa M. Valsta. 2008. "Mapping geographical variation in obesity in Finland." *European Journal of Public Health* 18:637-643.

Lampard, Richard. 2007. "Is social mobility an echo of educational mobility? Parents' educations and occupations and their children's occupational attainment." *Sociological Research Online* 12:16.

Lasker, Gabriel W. 1954. "The question of physical selection of Mexican migrants to the USA." *Human Biology* 26:52-58.

Lasker, Gariel W., and C.G. Nicholas Mascie-Taylor. 1988. "The framework of migration studies." In *Biological Aspects of Human Migration.Cambridge Studies in Biological Anthropology*, edited by C.G

Nicholas Mascie-Taylor andGabriel W. Lasker, 1-13.Cambridge: Cambridge University Press.

Lasker, Gabriel W., and C.G. Nicholas Mascie-Taylor. 1989. "Effects of social class differences and social mobility on growth in height, weight and body mass index in a British cohort." *Annals of Human Biology* 16:1–8.

Lasker, Gabriel W., and C.G. Nicholas Mascie-Taylor. 1996. "Influence of social class on the correlation of stature of adult children with that of their mothers and fathers." *Journal of Biosocial Science* 28:117-122.

Lewis, Sarah A., and John R. Britton. 1998. "Measles infection, measles vaccination and the effect of birth order in the aetiology of hay fever." *Clinical and Experimental Allergy* 28:1493–1500.

Li, Hongbin, Zhang, Junsen, and Yi Zhu. 2008. "The quantity-quality trade-off of children in a developing country: identification using Chinese twins." *Demography* 45:223-243.

Li, Jun, and Brent Taylor. 1993. "Factors affecting uptake of measles, mumps, and rubella immunization." *British Medical Journal* 307:168–171.

Li, Leah, Manor, Orly, and Chris Power. 2004. "Are inequalities in height narrowing? Comparing effects of social class on height in two generations." *Archives of Disease in Childhood* 89:1018-1023.

Li, Leah, and Chris Power. 2004. "Influences on childhood height: comparing two generations in the 1958 British birth cohort." *International Journal of Epidemiology* 33:1320-1328.

Machin, Stephen. 2004. "Education systems and intergenerational mobility." *Education* 1:1-26.

Malina, Robert M., Buschang, Peter H., Aronson,Wendy L., and Henry A. Selby. 1982. "Childhood growth status of eventual migrants and sedentes in a rural Zapotec community in the Valley of Oaxaca, Mexico." *Human Biology* 54:709–716.

Marjoribanks, Kevin. 1987. "Birth order and sibsize correlates of educational attainment and occupational status."*Psychological Reports* 61:147-150.

Mascie-Taylor, C.G. Nicholas. 1984. "The interaction between geographical and social mobility." In *Migration and Mobility, Biosocial Aspects of Human Movement. Symposia of the Society for the Study of Human Biology, vol. 23*, edited by Anthony J. Boyce, 161–178. London and Philadelphia: Taylor and Francis.

Mascie-Taylor, C.G. Nicholas. 1985. Biosocial correlates of IQ. In *The Biology of Intelligence*, edited by C.Turner, 99-127. Nafferton Books.

Mascie-Taylor,C.G. Nicholas, and Jasper L. Boldsen. 1985. "Regional and social analysis of height variation in a contemporary British sample." *Annals of Human Biology* 12:315-324.

Mascie-Taylor, C.G. Nicholas, andJohn B. Gibson. 1978. "Social mobility and IQ components." *Journal of Biosocial Science* 10:263-276.

Mascie-Taylor, C.G. Nicholas, and Gabriel G. Lasker. 1987. "Migration and changes in ABO and Rh blood group clines in Britain." *Human Biology* 59:337-344.

Mascie-Taylor, C.G. Nicholas, and Michael A. Little. 2004. "History of migration studies in biological anthropology." *American Journal of Human Biology* 16:365-378.

Méjean, Caroline, Traissac, Pierre, Eymard-Duvernay, Sabrina, El Ati Jalila, Delpeuch, Francis, and Bernard Maire. 2007. "Influence of socio-economic and lifestyle factors on overweight and nutrition-related diseases among Tunisian migrants versus non-migrant Tunisians and French." *BMC Public Health* 7:265.

Modin, Bitte. 2002. "Birth order and mortality: A life-long follow-up of 14,200 boys and girls born in early 20th century Sweden." *Social Science and Medicine* 54:1051–1064.

Mourant, Arthur E., Kopec, Ada C., and Kazimiera Domaniewska-Sobczak. 1976. *The Distribution of the Human Blood Groups and Other Polymorphisms*. Oxford, Oxford University Press.

Müller, Walter, and Yossi Shavit. 1998. "The institutional embeddedness of the stratification process: a comparative study of qualifications and occupations in thirteen countries." In *From School to Work: A Comparative Study of Educational Qualifications and Occupational Destinations*, edited by Yossi Shavit, and Walter Müller, 1–48. New York: Clarendon Press.

Nagel, Gabriele, Wabitsch, Martin, Galm, Christoph, Berg, Swantje, Brandstetter, Susanne, Fritz, Michael, Klenk, Jochen, Peter, Richard, Prokopchuk, Dmytro, Steiner, Ronald, Stroth, Sanna, Wartha, Olivia, Weiland, Stephan K., and Juergen Steinacker. 2009. "Determinants of obesity in the Ulm Research on Metabolism, Exercise and Lifestyle in Children (URMEL-ICE)." *European Journal of Pediatrics* 168:1259–1267.

Nettle, Daniel. 2003. "Intelligence and class mobility in the British population." *British Journal of Psychology* 94:551-561.

Park, Julie, Myers, Dowell, Kao, Dennis, and Min SeongHee. 2009. "Immigrant obesity and unhealthy assimilation: Alternative estimates of convergence or divergence, 1995-2005." *Social Science and Medicine* 69:1625-1633.

Parr, Nick. 2006. "Do children from small families do better?" *Journal of Population Research* 23:1-25.

Paulhus, Delroy L., Trapnell, Paul D., and David Chen. 1999. "Birth order effects on personality andachievement within families." *Psychological Science* 10:482-488.

Peytremann-Bridevaux, Isabelle, and Brigitte Santos-Egglma. 2007. "Use of preventive services of overweight and obese Europeans aged 50-79 years." *Journal of General Internal Medicine* 22:923-929.

Power, Chris, Manor, Orly, and Leah Li. 2002. "Are inequalities in height underestimated by adult social position? Effects of changing social structure and height selection in a cohort study." *British Medical Journal* 325:131-134.

Rasmussen, Finn, Johansson, Malin, and Henriette O. Hansen. 1999. "Trends in overweight and obesity among 18-year-old males in Sweden between 1971 and 1995." *Acta Paediatrica* 88:431-437.

Reeder BA, Chen Y, MacDonald SM, Angel A, and L. Sweet(1997) Regional and rural-urban differences in obesity in Canada. Canadian Heart Health Surveys Research Group. *Canadian Medical Association Journal* 157 (Suppl.1):S10-S16.

Rennie, Kirsten L., and Susan A. Jebb. 2005. "Prevalence of obesity in Great Britain." *Obesity Reviews* 6:11–12.

Ringdal, Kristen. 1993. "Migration and status attainment among Norwegian men." *Acta Sociologica* 36:327-342.

Rohde, Percy A., Atzwanger, Klaus, Butovskaya, Marina, Lampert, Ada, Mysterud, Iver, Sanchez-Andres, Angeles, and Frank J. Sulloway. 2003. "Perceived parental favoritism, closeness to kin, and the rebel of the family:The effects of birth order and sex." *Evolution and Human Behavior* 24:261-276.

Rona, Roberto J., and Douglas G. Altman. 1977. "National Study of Health and Growth: Standards of attained height, weight and triceps skinfold in English children 5 to 11 years old." *Annals of Human Biology* 4:501-523.

Rosenbaum, Stella, Skinner, Rosalind K., Knight, Ian B., and John S. Garrow. 1985. "A survey of heights and weights of adults in Great Britain, 1980." *Annals of Human Biology* 12:115-127.

Saunders, Peter. 1997. "Social mobility in Britain: an empirical evaluation of two competing explanations." *Sociology* 31:261–88.

Saunders, Peter. 2002. "Reflections on the meritocracy debate in Britain: A response to Richard Breen and John Goldthorpe." *British Journal of Sociology* 53:559–574.

Savage, Mike. 1988. "The missing link? The relationship between spatial mobility and social mobility." *British Journal of Sociology* 39:554-577.

Scarborough, Peter, and Steven Allender. 2008. "The North-South gap in overweight and obesity in England." *British Journal of Nutrition* 100:677-684.

Schober, Edith, Rami, Birgit, Kirchengast, Sylvia, Waldhör, Thomas, and Reinhart Sefranek. 2007. "Recent trend in overweight and obesity in male adolescents in Austria: a population-based study." *European Journal of Pediatrics* 166:709–714.

Sen, Anindya, and Anthony Clemente. 2010. "Intergenerational correlations in educational attainment: birth order and family size effects using Canadian data." *Economics of Education Review* 29;147-155.

Shavit, Yossi, and Hans-Peter Blossfeld. 1993. *Persistent inequalities: a comparative study of educational attainment in thirteen countries.* Boulder Colorado: Westview Press.

Shields, Margot, and Michael Tjepkema. 2006. "Regional differences in obesity." *Health Reports* 17:61-67.

Shin, Heeju. 2007. "Child health in Peru: importance of regional variation and community effects on children's height and weight." *Journal of Health and Social Behavior* 48:418-433.

Sieben, Inge, and Paul M. De Graaf. 2001. "Testing the modernization hypothesis and the socialist ideology hypothesis: a comparative sibling analysis of educational attainment and occupational status." *British Journal of Sociology* 52:441 - 467.

Singh, L.P., and G.A. Harrison. 1996. "Impact of migration, environment and socioeconomic conditions on the physique of Sikhs." *Journal of Biosocial Science* 28:101–116.

Singh-Manoux, Archana, Gourmelen, Julie, Ferrie, Jane, Silventoinen, Karri, Gueguen, Alice, Stringhini, Silvia, Nabi, Hermann, and Mika Kivimaki. 2010. "Trends in the association between height and socioeconomic indicators in France, 1970–2003." *Economics and Human Biology* 8:396-404.

Smith, George D., Hart, Carole, Hole, David, MacKinnon, Pauline, Gillis, Charles, Watt Graham C., Blane, David, and Victor Hawthorne. 1998. "Education and occupational social class: which is the more important indicator of mortality risk?" *Journal of Epidemiology and Community Health* 52:153–160.

Sorjonen, Kimmo, Hemmingsson, Tomas, Lundin, Andreas, and Bo Melin. 2011. "How social position of origin relates to intelligence and level of education when adjusting for attained social position." *Scandinavian Journal of Psychology* 52:277–281.

Steegmann, A. Theodore. 1985. "18th century British military stature: growth cessation, selective recruiting, secular trends, nutrition at birth, cold and occupation." *Human Biology* 57: 77–95.

Steelman, Lala C., and Brian Powell. 1985. "The social and academic consequences of birth order: real, artifactual, or both?" *Journal of Marriage and Family* 47:117-124.

Strachan, David P., Rudnicka, Alicja R., Power, Chris, Shepherd, Peter, Fuller, Elizabeth, Davis, Adrian, Gibb, Ian, Kumari, Meena, Rumley, Ann, Macfarlane, Gary J., Rahi, Jugnoo, Rodgers, Bryan, and Stephen Stansfeld. 2007. "Lifecourse influences on health among British adults: Effects of region of residence in childhood and adulthood." *International Journal of Epidemiology* 36:522-531.

Sulloway, Frank J. 1996. *Born to rebel: Birth order, family dynamics, and creative lives*. New York: Pantheon.

Szklarska, Alicja, Lipowicz, Anna, Łopuszańska, Monika, Bielicki, Tadeusz, and Sławomir Kozieł. 2008. "Biological condition of adult migrants and nonmigrants in Wrocław, Poland." *American Journal of Human Biology* 20:139-145.

Terrell, Thomas R., and C. G. Nicholas Mascie-Taylor. 1991. "Biosocial correlates of stature in a 16-year old British cohort." *Journal of Biosocial Science* 23:401-408.

Tomasson, Richard F. 1966. "Social mobility and family size in two high-status populations." *Eugenics Quarterly* 13:113-121.

Van Bavel, Jan, Moreels, Sarah, Van de Putte, Bart, and Koen Matthijs. 2011. "Family size and intergenerational social mobility during the fertility transition." *Demographic Research* 24:313-344.

Von Stumm, Sophie, Gale, Catharine R., Batty, David G., and Ian J. Deary. 2009. "Childhood intelligence, locus of control and behaviour disturbance as determinants of intergenerational social mobility: British Cohort Study 1970." *Intelligence* 37:329-340.

Von Stumm, Sophie, Macintyre, Sally, Batty, David G., Clark, Heather, and Ian J. Deary. 2010. "Intelligence, social class of origin, childhood behavior disturbance and education as predictors of status attainment in midlife in men: The Aberdeen Children of the 1950s study." *Intelligence* 38:202-211.

Waldhör, Thomas, Schober, Edith, Rami, Birgit, and the Austrian Diabetes Incidence Study Group. 2003. "Regional distribution of risk for childhood diabetes in Austria and possible association with body mass index." *European Journal of Pediatrics* 162:380–384.

Waller, Jerome H. 1971. "Achievement and social mobility: relationships among IQ score, education and occupation in two generations." *Social Biology* 18:252-259.

Wheeler, Erica, and Swee P. Tan. 1983. "Trends in the growth of ethnic Chinese children living in London." *Annals of Human Biology* 10:441–446.

Will, Beata, Zeeb, Hajo, and Bernhard T. Baune. 2005. "Overweight and obesity at school entry among migrant and German children: a cross-sectional study." *BMC Public Health* 5:45.

Zajonc, R. B. 1976. "Family configuration and intelligence." *Science* 192:227–236.

Zajonc, R. B., and Gregory B. Markus. 1975. "Birth order and intellectual development." *Psychological Review* 82:74–88.

Zielińska, Dorota. 1991. "Is there a selective rural–urban migration in respect to height and weight?" *American Journal of Human Biology* 3:363–368.

In: Biological Implications of Human Mobility ISBN: 978-1-63485-644-7
Editors: Slawomir Koziel et al.

Chapter 9

POPULATION DYNAMICS, DIETARY HABITS AND LIFESTYLE CHANGES IN THE EASTERN ADRIATIC: A REVIEW

***Saša Missoni*[1,2] *and Jelena Šarac*[1,*]**
[1]Institute for Anthropological Research, Zagreb, Croatia
[2]"Josip Juraj Strossmayer" University of Osijek, School of Medicine, Osijek, Croatia

ABSTRACT

A general shift in dietary habits and lifestyle practices has been observed in Croatia and such relatively abrupt changes have had an especially significant impact on the Eastern Adriatic islands. Croatian insular populations, known for their traditional Mediterranean diet, currently fit within the global trend of transition, globalisation and urbanisation, characterized by an increased intake of high-energy products, decreased level of physical activity and emigration from rural into urban areas. This negative trend is also associated with the development of pathogenic conditions and complex diseases, such as metabolic syndrome, obesity, diabetes mellitus and hypertension, conditions which represent a major public health problem nowadays.

* Corresponding Author address: Insitute for Anthropological Research, Gajeva 32, 10000 Zagreb. Email: jelena.sarac@inantro.hr.

Keywords: dietary habits, lifestyle, health, Eastern Adriatic, Croatia

INTRODUCTION

Republic of Croatia is divided into two basic geographic and climatic regions with significant differences in lifestyle and dietary patterns - a continental and a Mediterranean (coastal and insular) region. Despite the regional differences, a general shift in nutritional habits and lifestyle has been observed in the direction of globalization and transition in both regions. Such relatively abrupt changes have had an especially significant impact on the insular Eastern Adriatic populations. Croatian islands communities, known for their traditional Mediterranean diet, are most likely to disappear in the near future because of the ongoing globalization and depopulation process, and because of the newly introduced changes in lifestyle. A notable influence of "global" diet has been noticed in the Mediterranean region, which has consequently led to the increased prevalence of hypertension, metabolic syndrome, diabetes mellitus, coronary heart diseases etc. on Croatian islands. This kind of research has already been conducted in ex-Yugoslavia in the 1970ies, as a part of The Seven Countries Study (Keys et al. 1981) and it is a powerful tool for studying the influence of dietary and other transitional changes on phenotypic traits in a population.

POPULATION DYNAMICS OF THE EASTERN ADRIATIC

Croatia's Eastern Adriatic Sea coastline is 1 777,3 kilometers long and as such represents one of the most indented coastlines in the world. Although the number of Croatian Adriatic islands, islets and reefs is 1 246, only 47 are inhabited, with altogether 121,625 inhabitants on all islands (<3% of the total Croatian population according to the 2011 census). The population peak in the Eastern Adriatic was in 1921 at 173,503 and since then the number of inhabitants in Croatian island communities is in a systematic decline. Also, the Croatian island population is older than the general Croatian population, with almost 30% of the island population aged 60 and over (according to the 2011 census). A slight and short expansion period was observed in the early 1980s, as a result of both 'pull' factors, in terms of the rapid growth of tourism and, hence, raising standard of living and increased employment opportunities and 'push' factors, in terms of economic crisis in the urban areas (Starc and Stubbs 2013).

The populations of the Eastern Adriatic islands are represented by a range of distinct demographic histories according to the founding times, origin and number of founders, bottleneck and admixture events, length of isolation, and historical fluctuations in population size, but they have all been subject to the same major migratory and (pre)historical events. The earliest evidence of human occupation of Adriatic islands has been dated to Upper Paleolithic and Neolithic period and the first historically known people that have lived on this island were Illyrians, a heterogeneous group of 'tribes' who spoke an Indo-European language (Katičić 1964). Archeological findings suggest that Eastern Adriatic islands, with their strategic location on ancient maritime routes, have played an important role in Mediterranean trading networks in prehistory (Forenbaher and Miracle 2005). With the Greek colonization (7th century BC) the Eastern Adriatic gradually entered the realm of history, followed by the rule of the Roman and subsequently Byzantine Empire. The first well-documented historic and demographic event that had left significant impact on the population structure of Croatia was the coming of Slavs. Slavic expansion into Southeastern Europe (SEE) and Croatia started as early as 3th century and ended with a big migration wave in 6th century after which the current population landscape of this region gradually started to shape (Curta 2006). They colonized the continental part of the region first and the islands were populated somewhat later. The arrival of the new population marked the beginning of a new historical, political, ethnic and genetic grouping and of a turbulent period in the history of Eastern Adriatic islands - Croatian kings and dukes, Venetians, Austro-Hungarian Monarchy, Italy and Yugoslavia have reigned over contemporary Croatian territory for centuries. The following important migrational period occurred between the 15th and the 18th century, in a time characterized by Turkish expansion into SEE and the Balkan Peninsula. Invasion of savage Turkish troops on Croatia resulted in great population movements across the continental part of the region and also in significant migrations from mainland to the Adriatic islands (Stavrianos and Stoianovich 2000, Rudan et al. 2004 and the references therein). In the early 20th century, the Austro-Hungarian Empire was replaced by Italy in the North and the Republic of Yugoslavia in the South and a rapid emigration trend and depopulation process was initiated, starting with 1921 and ending only in the last decade of the century. It was caused by both the economic factors, such as the agrarian overpopulation, land industrialization, deagrarization, the crises of certain agricultural activities, monoculture on some of the islands, etc., and also by political reasons in form of political emigration. Depopulation due to emigration was crucial in creating a general depopulation trend up to the 1960s, and after that period it was joined by the natural depopulation as well, leading to extreme conditions in many island communities (Lajić and Mišetić 2006). Not surprisingly, the island development was also not among the political priorities in the 1990ies, in the context of the more pressing issues in terms of Croatia's independence war (1991-1995).

Although migration as such is not a new phenomenon, its character has, however, fundamentally changed in the last century due to revolutionary technological and infrastructural developments and the incorporation of regions and countries across the globe within international capitalism. The major advances in transport and communication technology, improvements in knowledge flows, a perceived lack of local economic opportunities, and growing level of welfare in the second half of the twentieth century have facilitated, accelerated and transformed the globalization process on the Croatian islands and increased the capabilities and aspirations of people from island communities to migrate, maintain transnational ties and construct transnational identities. Current emigration process in the Eastern Adriatic cannot be characterized as a desperate response to destitution or a last resort to escape from extreme conditions of poverty and unemployment. In line with the new economics of labour migration and livelihood approaches, 'modern' migrations are rather a deliberate attempt of mostly younger generations to improve their social and economic status and, hence, to overcome local development constraints by moving to more urban areas in search for employment opportunities. Though many of the islands once were self-sustained due to agriculture and/or fishing, increased globalisation and urbanisation has put them at a disadvantage as the importance of traditional activities has declined. They are instead shifting more and more from agriculture and fishery towards a more lucrative and physically less demanding branch of economy- tourism, making it an active generator of the current economic development on Croatian islands.

Molecular-Genetic and Genetic-Epidemiological Research in the Eastern Adriatic

Since the beginning of anthropological research in Croatia, Eastern Adriatic islands have been a focus of diverse and wide-ranging research. They represent well-characterized genetic isolates, concerning their ethnohistory, biological trait measurements, disease prevalence, migration patterns and environmental and sociocultural characteristics. The results of numerous studies (Rudan et al. 2004 and the references therein) indicate that village populations on Croatian islands have preserved certain genetic specifics over the course of history to the present day, and measures of kinship and genetic distances revealed isolation of such communities from each other and from the mainland. Since such small communities can reflect large demographic processes that happened in human prehistory and history, the scientific value of genetic isolates has also been proven in the field of population genetics and archeogenetics.

Due to their environmental homogeneity and common practice of endogamy, Croatian isolates have also been recognized as ideal tool for mapping Mendelian disease traits (Saftić et al. 2006), finding candidate genes involved in the development of many complex diseases (Deka et al. 2008, 2012, Rudan et al. 1999, 2003) and for conducting linkage and association analyses such as GWAS (e.g., Vitart et al. 2008, Barbalić et al. 2009, Karns et al. 2012, 2013). Expression of genes responsible for the occurrence of complex diseases is often enhanced in such isolated populations, which is why genetic-epidemiological research in population isolates is of great importance, giving us additional insight into the interplay between intrinsic (genes) and extrinsic risk factors (diet) for disease development. One such genome-wide analysis of 1,300 inhabitants has been performed on the island of Hvar and the results detected new SNP variants associated with cardiometabolic traits, but also with uric acid levels and body height. The range of performed studies on Croatian island isolates emphasize their potential influence on public health through the investigation of the links between genetic variation and specific health risk factors and through the novel insight into the disease mechanisms and components based on the studies of isolated communities (Rudan et al. 2004, Peltonen et al. 2000, Missoni et al. 2012). Illustrative examples of specific epidemiological settings in such island communities are well described in some of the Eastern Adriatic isolates, which are characterized by specific genetic disorders and conditions, such as dwarfism on Krk, "Mal de Meleda" on Mljet, extreme consanguinity and learning disability on Susak and population genetics of cancer on Brač, Hvar, Korčula, Vis and Lastovo etc. (Rudan and Rudan 2004).

DIETARY HABITS AND NUTRITIONAL STUDIES IN CROATIA

Republic of Croatia is divided into two basic geographic and climatic regions with significant differences in lifestyle and dietary patterns - a continental and a Mediterranean (coastal and insular) region. Continental Croatia is characterized by a shorter vegetation period and consequently also by a reduced intake of fresh fruits and vegetables throughout the year, when compared to the Mediterranean part of the country. The most significant difference in dietary habits between the regions is in the proportion of saturated and unsaturated fatty acids in the diet and in the general calorie intake. Higher intake of saturated fatty acids of animal origin is more common in continental Croatia, together with the increased consumption of smoked meat products. Also, typical preparation of food there is roasting or deep-fat frying. Diet of farmers in the coastal part of the country and on the islands can on the other hand be seen as a variant of the Mediterranean diet, characterized mostly by the consumption of brown bread, olive oil, red wine, fruits and vegetables. Such divergence of dietary habits in one population is potentially interesting for studying the influence of diet and globalization trends on phenotypic traits and health in a population.

Studying the relationship between diet and etiology of various complex, chronic diseases has a long tradition in Croatia. In 1958, Buzina and collaborators started a long-term and complex research in Slavonia (continental Croatia) and Dalmatia (Mediterranean Croatia), as a part of the famous Keys's Seven Countries Study (Buzina et al. 1970, Keys et al. 1981). Results showed that the biggest differences in dietary habits between these two regions were in the use of fat, according to which the ratio of unsaturated and saturated fatty acids was 1:1 in Slavonia and 3:1 in Dalmatia, while the content of mono-, di- and tri unsaturated fatty acids was bigger in Dalmatia. Also, more patients with hypertension were found in Slavonia. It was also proved that higher socioeconomic status was associated with higher weight, higher blood pressure and higher level of cholesterol in blood (Buzina et al. 1970). Another nutritional study conducted in these two distinct Croatian regions was performed by Kaić-Rak and collaborators in 1980s. They investigated the onset and development of stomach cancer in association with dietary habits and concluded that the incidence of stomach cancer in the continental Croatia is greater than in the southern, Mediterranean parts of the country. The study also showed that a more qualitative nourishment is related to the island populations, where a more satisfactory balance of saturated and unsaturated fats is observed, They mostly use vegetable oils (especially olive oil), while 90% of fat consumption in the continental region are fats of animal origin. It was also discovered that the consumption of complex carbohydrates, vegetables and fruit was higher on the islands, leading to a higher intake of vitamins A and C as important elements in the prevention of stomach cancer (Kaić-Rak et al. 2007). In the 1990ies, Rudan and collaborators discovered differences in the rate of incidence of stomach, colon and pancreas cancer between the two regions in Croatia, and their results showed that the Mediterranean diet decreases the risk of stomach and pancreas cancer, and serves as a protective factor in the appearance of atherosclerosis (Rudan et al. 2003).

LIFESTYLE CHANGES AND OBSERVED TRENDS ON CROATIAN ISLANDS

Considerable number of studies has in the last years been focused on studying the influence of changes in dietary habits and lifestyle on the development of obesity and metabolic syndrome in Croatia. Research conducted in the 90s on the island of Hvar already showed a tendency towards a change in dietary habits of the contemporary population, including higher intake of meat, pasta, cakes and coffee (Smolej-Narančić 2000). In the last ten years, results of research performed on other Croatian island populations (Missoni 2009) have shown similar trends. Although the diet on the islands is predominantly Mediterranean, a significant increase in the consumption of red meat, poultry, milk, dairy products, sugar and industrial products and a decreased consumption of fish, fruits and vegetables has been observed. This has had a significant effect on the balance between saturated and unsaturated fatty acids, increasing the levels of saturated ones due to intake of red meat and industrial products (Missoni 2009). Intake of red meat and poultry, fruits and vegetables is in line with the national average, while the consumption of other meat is 2–3 times per week. Intake of fish is higher than the national average, but still lower than in other Mediterranean countries. The study conducted by Kolčić et al. (2006) is also worth mentioning, since it has established a high prevalence of complex diseases on Croatian island isolates. The highest prevalence of metabolic syndrome has been observed on the island of Mljet, according to WHO (53%). A study by Deka et al. (2012) has indicated a high prevalence of the syndrome on the island of Hvar – 25% according to NCEP and 38% according to IDF. One of the highest prevalence of obesity in Croatia has been recorded on the island of Vis, according to all three measures for obesity detection (body mass index, waist circumference and waist-to-hip circumference ratio). Also, high results have been obtained for hypertension, metabolic syndrome, diabetes mellitus, coronary heart diseases and cerebral stroke on the same island (Missoni 2009).

Such rapid changes in lifestyle and dietary habits observed in the last few decades can be explained from several perspectives. The processes of globalization and transition in Croatia can be considered as the most responsible for the abrupt and drastic changes in dietary habits and lifestyle on Croatian islands. Hard physical labour is being increasingly replaced with mechanisation and traditional agriculture and fishery with a physically less demanding tourism (e.g., current wine-growing economy amounts for only 8% of the same economy before WWII) (Missoni 2012). However, almost exclusive orientation to summer tourism brings Croatian islands to a new serious problem - intense touristic demand occurring in a short period of time puts great pressure upon the island's socio-economic and eco-system and this problem is yet to be resolved (Petrić and Mikulić 2013). Dramatic transitional changes also occurred in the labor structure on the Croatian islands, primarily involving a dismissal of workers in numerous state owned enterprises and increase in the number of unemployed, increase in employment in the private sector due to the privatization of the central planned economy, a drastic decrease of people working in the agricultural sector and the emergence of a new managerial elite and a new middle class (Cerami 2009). One key feature of the globalisation phenomenon is the neglection of agriculture and the appearance of supermarkets, directing people towards consumption of cheaper, more easily accessible, high-energy and nutrient-poor food. Such diet, which includes more saturated fatty acids, cholesterol and simple carbohydrates and not enough fruits and vegetables, fibres, vitamins, minerals etc., consequently has considerable effect on the development and distribution of obesity and metabolic syndrome. Our insular populations fit nicely within this global trends with increased intake of high-energy products and decreased level of physical activity, as already suggested in the WHO report (WHO 2003).

Another theory holds that in the past, fish and brown bread have been perceived as symbols of poverty, while red meat and white bread were considered elite food. Nowadays, when these "elite-products" are equally available, people from Croatian insular, rural regions turn to them more often.

Accordingly, studies have shown that wealthier people tend to consume more red meat, poultry and refined sugar (Missoni 2012).

As far as is known, changes in dietary habits and lifestyle in general have never been introduced in human populations so fast. The sudden and swift shift in dietary habits on the islands accompanied by a drastically reduced level of physical activity has been observed as a strong trigger for development of diverse pathogenic changes, general obesity trend, and occurrence of metabolic syndrome, diabetes mellitus and coronary heart diseases (Leonard 2012, Janković 2013). We can therefore propose an additional hypothesis that populations have not yet adapted to the new habits and food products that are produced and processed in a different way. Although it is still not scientifically proven, we cannot neglect the existence of new and complex chemical compounds that are affecting the products of our ecosystem (their growth and processing) and whose effect on human health is questionable (Missoni 2012).

Future Research and Strategy

Given the high prevalence of complex diseases associated with the changes in dietary habits and lifestyle, the need for a return to the traditional Mediterranean cuisine in the island communities is raised. Prevention in form of systematic education, public symposiums and workshops in schools and medical centres should be a priority, together with further anthropological studies investigating island communities in Croatia. Such insular populations are a powerful tool for elucidating the interplay between intrinsic (genes) and extrinsic factors (diet) in a population and genetic-epidemiological research conducted in such communities can be seen as great contribution to the improvement of the general public health system. Another interesting perspective would be the study of people that originate from the Croatian islands, but live in other parts of the country. We would then have one population that has been exposed to different extrinsic factors, such as different dietary habits and this could offer interesting new insights. Longitudinal research conducted on children and adolescents would also be of significant value, since it would enable us to study the development of obesity and metabolic syndrome and give us a possibility to detect possible triggers for the onset of such diseases and ways to prevent them. In general, an interdisciplinary approach to such matters is crucial in future research – an

interaction between the sociocultural and biomedical research, together with an evolutionary perspective on lifestyle changes and population dynamics.

References

Barbalić, Maja, Narančić, Nina S., Škarić-Jurić, Tatjana, Salihović, Marijana P., Klarić, Irena M., Lauc, Barać L., Janićijević, Branka, Farrall, Martin, Rudan, Igor, Campbell, Harry, Wright, Alan F., Hastie, Nicholas D., and Pavao Rudan. 2009. "A quantitative trait locus for SBP maps near KCNB1 and PTGIS in a population isolate". *American Journal of Hypertens* 22:663–668.

Buzina, Ratko, Keys, Ancel, Mohaček, Ivan, Marinković, M., Hahn, Arpad, and Henry Blackburn. 1970. "Five year follow up in Dalmatia and Slavonia". In *Coronary heart disease in seven countries*, edited by Ancel Keys. New York: The American Heart Association.

Cerami, Alfio. 2009. "Socio-Economic Transformations in Post-Communist Countries: Central and Eastern Europe and Russia Compared". *Romanian Journal of Political Sciences* 9(2):25-41.

Curta, Florian. 2006. "Southeastern Europe in the Middle Ages". Medieval Textbooks series. Cambridge: Cambridge University Press.

Deka, Ranjan, Duraković, Zijad, Niu, Weimin, Zhang, Guangcheng, Karns, Rebekah, Smolej-Narančić, Nina, Missoni, Sasa., Carić, Damira, Carić, Tomislav, Rudan, Dusko, Salzer, Branka, Chakraborty, Ranajit, and Pavao Rudan. 2012. "Prevalence of metabolic syndrome and related metabolic traits in an island population of the Adriatic". *Annals of Human Biology* 39:46–53.

Deka, Ranjan, Smolej-Narančić, Nina, Xi Huifeng, Turek, Stjepan, Cubrilo-Turek, Mirjana, Vrhovski-Hebrang, Danijela, Janićijević, Branka, Tomljenović, Andrea, Szirovicza, Lajos, Jin, Li, Chakraborty, Ranajit, and Pavao Rudan. 2008. "Metabolic syndrome in an island population of the eastern Adriatic coast of Croatia". *Collegium Antropologicum* 32:85–91.

Forenbaher, Staso, and Preston T. Miracle. 2005. "The spread of farming in the Eastern Adriatic". *Documenta Praehistorica* 33:89-100.

Janković, Ivor. 2013. "Evolucija hominina: biološka i kulturna adaptacija s posebnim osvrtom na prehranu". (In Croatian) Paper presented at the 2. interdisciplinary scientific symposium Croatian Medical Association, Zagreb, Croatia.

Kaić-Rak, Antoinette, Pucarin-Cvetković, Jasna, and Ignac Kulier. 2007. "Dietary habits: Croatian health survey". *Acta Medica Croatica* 61: 259-265.

Karns, Rebekah, Succop, Paul, Zhang, Ge, Sun, Guangyun, Indugula, Subba R., Havas-Augustin, Dubravka, Novokmet, Natalija, Durakovic, Zijad, Milanovic, Sanja M., Missoni, Sasa, Vuletic, Silvije, Chakraborty, Ranajit, Rudan, Pavao, and Ranjan Deka. 2013. "Modeling metabolic syndrome through structural equations of metabolic traits, comorbid diseases, and GWAS variants". *Obesity* 21:E745-54.

Karns, Rebekah, Zhang, Ge, Sun, Guangyun., Indugula, Subba R., Cheng, Hong, Havas-Augustin, Dubravka, Novokmet, Natalija, Rudan, Dusko, Durakovic, Zijad, Missoni, Sasa, Chakraborty, Ranajit, Rudan, Pavao, and Ranjan Deka. 2012. "Genome-wide association of serum uric acid concentration: replication of sequence variants in an island population of the Adriatic coast of Croatia". *Annals of Human Genetics* 76:121–127.

Katičić, Radoslav. 1964. "Language of the autochtonous population of Illyrian provinces". Paper presented at the Symposium on the Illyrian territorial and chronological boundaries in prehistory, Sarajevo: ANUBiH.

Keys, Ancel, Aravanis, Christ, Van Buchem, Frans S.P., Blackburn, Henry, Buzina, Ratko, Đorđević, Boban S., Donats, Anastasios S., Fidanza, Flaminio, Karvonen, Mariti J., Kimura, Naboru, Menotti, Alessandro, Nedeljković, Srecko, Puddu, Vittorio, Punsar, Sven, and Henry L. Taylor. 1981. "The diet and all-causes death rate in the seven countries study". *The Lancet* 8237:58-63.

Kolčić, Ivana, Vorko-Jović, Ariana, Salzer, Branka, Smoljanović, Mladen, Kern, Josipa, and Silvije Vuletić. 2006. "Metabolic syndrome in a metapopulation of Croatian island isolates". *Croatian Medical Journal* 47:585–592.

Lajić, Ivan, and Roko Mišetić. 2007. "Otočni logaritam. Aktualno stanje i suvremeni demografski procesi na jadranskim otocima". (In Croatian) Zagreb: Institute for Migration and Ethnic Studies.

Leonard, William R. 2012. "Human Nutritional Evolution". In *Human Biology: An evolutionary and biocultural perspective*, edited by Sara Stinson, Barry Bogin, and Denis O'Rourke. Hoboken, NJ: John Wiley and Sons, Inc.

Missoni, Sasa. 2006. "Nutritional studies in Croatia--a century of research". *Collegium Antropologicum* 30: 673-96.

Missoni, Sasa. 2009. "Međuodnos prehrambenih navika i kompleksnih fenotipskih svojstava stanovništva otoka Visa". (In Croatian) Zagreb: Croatian Anthropological Society.

Peltonen, Leena, Palotie, Aarno, and Kenneth Lange. 2000. "Use of population isolates for mapping complex traits". *Nature Review Genetetics* 1:182-190.

Petrić, Lidija, and Davorka Mikulić. 2013. "Protected areas and tourism development on Croatian islands: coexistence or divergence?". In *Island Sustainability II*, edited by Srecko Favro and Carlos A. Brebbia. UK: WIT Press.

Rudan, Igor, Ranzani, Guglielmina N., Strnad, Marija, Vorko-Jović, Ariana, John, V., Unusić, Josi, Kern, Josipa, Ivanković, Davor, Stevanović, Ranko, Vuletić, Silvije, and Pavao Rudan. 1999. "Surname as "cancer risk" in extreme isolates: example from the island of Lastovo, Croatia". *Collegium Antropologicum* 23:557–569.

Rudan, Igor, Rudan, Diana, Campbell, Harry, Carothers, Andrew, Wright, Alan, Smolej-Narančić, Nina, Janićijević, Branka, Jin, Li, Chakraborty, Ranajit, Deka, Ranjan, and Pavao Rudan. 2003. "Inbreeding and risk of late onset complex disease". *Journal of Medical Geneticts* 40:925–932.

Rudan, Pavao, Janićijević, Branka, Jovanović, Veljko, Miličić, Jasna, Narančić, Nina S., Sujoldžić, Anita, Szirovicza, Lajos, Škarić-Jurić, Tatjana, Barać Lauc, Lovorka, Lauc, Tomislav, Klarić, Irena M., Peričić, Marijana, Rudan, Diana, and Igor Rudan. 2004. "Holistic anthropological research of Hvar Islanders, Croatia--from parish registries to DNA studies in 33 years". *Collegium Antropologicum* 28(Suppl 2):321–343.

Rudan, Igor, and Pavao Rudan. 2004. "From genomic advances to public health benefits: the unbearable lightness of being stuck". *Collegium Antropologicum* 28:483-507.

Saftić, Vanja, Rudan, Diana, and Lina Zgaga. 2006. "Mendelian diseases and conditions in Croatian island populations: historic records and new insights". *Croatian Medical Journal* 47:543–552.

Smolej-Narančić, Nina. 2000. "Physical fitness and lifestyle in a Mediterranean setting". Presented at the 5th International Congress on Physiological Anthropology, edited by Sohn and Jang-Yeul. Seul: The Korean Society of Living Environmental System.

Starc, Nenad, and Paul Stubbs. 2013. "Tangled up in blue: the Croatian islands and participatory development planning". In *Island Sustainability II*, edited by Srecko Favro and Carlos A. Brebbia. UK: WIT Press.

Stavrianos, Leften, and Traian Stoianovich. 2000. *The Balkans Since 1453*. NY: New York University Press.

Turek, Stefan, Rudan, Igor., Smolej-Narančić, Nina, Szirovicza, Lajos, Čubrilo-Turek, Mirjana, Zerjavić-Hrabak, Vlasta, Rak-Kaić, Antoinette, Vrhovski-Hebrang, Danijela, Prebeg, Zivka., Ljubičić, Mate, Janićijević, Branka, and Pavao Rudan. 2001. "A large cross-sectional study of health attitudes, knowledge, behaviour and risks in the post-war Croatian population (the First Croatian Health Project)". *Collegium Antropologicum* 25:77-96.

Vitart, Veronique, Biloglav, Zrinka, Hayward, Caroline, Janićijević, Branka, Smolej-Narančić, Nina, Barać, Lovorka, Peričić, Marijana, Klarić, Irena M., Škarić-Jurić, Tatjana, Barbalić, Maja, Polašek, Ozren, Kolčić, Ivana, Carothers, Andrew, Rudan, Pavao, Hastie, Nick, Wright, Alan, Campbell, Harry, and Igor Rudan. 2006. "3000 years of solitude: extreme differentiation in the island isolates of Dalmatia, Croatia". *European Journal of Human Genetics* 14:478–487.

WHO/FAO expert consultation. 2003. "Diet, nutrition and the prevention of chronic disease". Geneva: WHO Press.

In: Biological Implications of Human Mobility ISBN: 978-1-63485-644-7
Editors: Slawomir Koziel et al.

Chapter 10

MIGRANTS IN THE CITY – INTERNATIONAL MIGRATION AND WOMEN'S REPRODUCTIVE HEALTH WITH SPECIAL RESPECT TO PUNJABI WOMEN IN VIENNA, AUSTRIA

Maryam Singh and Sylvia Kirchengast*
Department of Anthropology, University of Vienna, Austria

ABSTRACT

Migration has complex effects on immigrant health. Unfortunately there is still a lack of scientific analyses concerning the effects of migration on women's health, in particular on reproductive health. The current study aimed to compare reproductive health parameters of Indian women originating from Punjab who migrated to Vienna Austria with those of Indian women still living in Punjab. A series of 107 women ageing between 18 and 58 years (x = 36.1yrs; ±11.3) was enrolled in the present study.65 women lived in the district of Jalandhar in Punjab, 50 Punjabi women lived as migrants in Vienna Austria. Data collection was carried out by face to face interviews based on a structured questionnaire in Vienna Austria and in Punjab, India. Significant differences between Punjabi women in Austria and Punjabi women in India occurred in the

* Corresponding Author address: Department of Anthropology University of Vienna, Althanstrasse 14, A-1090 Vienna, Austria. Email: sylvia.kirchengast@univie.ac.at.

age at marriage, age at first birth and the duration of lactation. Educational level was the most important factor influencing reproductive health parameters among Indian women in Austria and in Punjab. With increasing educational level number of offspring decreased and the age at first birth increased. Educational level influenced the use of contraceptives, health related behaviour and the duration of lactation. Indian women as immigrants are still doubly disadvantaged. These women neither have a high status in their own Indian society because they are women, but they have also a low status within the Austrian society because they are migrants.

Keywords: reproductive health, socioeconomic factors, immigrant women, India, Austria

INTRODUCTION

Migration of individuals and whole populations is an ancient phenomenon and nothing new in the history of *Homo sapiens*. Up to the late 19th century migrations from rural to urban areas prevail. Recently people are moving not only in greater numbers than ever before but also over greater distances. Several push factors such as a lack of resources, poverty or war but also pull factors such as better educational or job opportunities or even marriage promote transnational migration worldwide. Today, more than 20 million persons in Western Europe are immigrants living in a host country which differs culturally markedly from their country of origin (Bollini et al. 2009, Deville et al. 2011). Furthermore the majority of these immigrants stem from rural areas and moved not only into a new country, they migrate into urban centres, which are especially attractive for immigrants. In these big cities however, most immigrants belong to disadvantaged groups living under especially unfavorable conditions. They are confronted with new cultural patterns, different eating and life style habits. Furthermore immigrants are a minority group in the host country, often with a lower socioeconomic position (SEP) than the host population. Minority status but also a low SEP is associated with increased chronic distress and as a consequence higher morbidity rates (Dotevall et al. 2000, Landman and Cruishank 2001, Bongard et al. 2002, Uitewaal et al. 2004a, 2004b, Daryani et al. 2005, Abate and Chandalia 2007, Misra and Ganda 2007, Adanu and Johnson 2009, Ujcic-Voortman et al. 2009). Consequentlyimmigrants' health is a pressing issue from a public health viewpoint. This is particularly true of women's health and reproductive health because it is estimated that women comprised 50% of migrants and the majority of them is of reproductive age or younger (Sargent and Larchanche 2011). Several surveys from the United Kingdom (Bundey et al. 1991), Belgium (Muynck 1997), the Netherlands (Jonkers et al. 2011), the United States (Madan et al. 2006, Uraquia et al. 2010) and Canada (Brotto et al. 2008) have shown similar results concerning female immigrant's reproductive health. Overall, as compared to women of the host population, immigrant women showed a clear disadvantage for various parameters of reproductive health: they showed a higher incidence of low birth weight, preterm delivery, perinatal mortality and congenital malformations (Bollini et al. 2009). Furthermore reproductive health behaviour and knowledge is poorer among immigrant women in comparison to the host population (Brotto et al. 2008). Immigrants' reproductive health behaviour however differs also from that typical of their country of origin. Research on immigrant fertility in Western Europe has shown a decline in fertility levels by duration of stay and by immigrant generation. Fertility levels seem to develop towards a convergence to the typical fertility level of the host population very fast (Schoenmaeckers et al. 1999; Fargues 2006; Milewski 2011). But not only fertility levels change after migration, a cohort study by Hawkins et al. (2008) revealed that female migrants in UK adopted adverse maternal behaviour such as smoking and alcohol drinking during pregnancy and a reduction of breast feeding very fast. Unfortunately there is still a lack of scientific analyses concerning the effects of migration or immigrant status on women's reproductive health (Bollini et al. 2009; Iglesias et al. 2003, Minnis 2010). This is especially true of small minority groups, such as Indian women in Austria. The Indian minority is a very small one in Austria and up to now no scientific analysis concerning reproductive health issues of Indian women in Austria was carried out. Information regarding reproductive health behaviour of Indian immigrant women is mainly available for Canada and UK (Brotto et al. 2008, Twamley et al. 2011). These studies implicate marked differences between Indian immigrant women and the host population concerning reproductive health and health related behaviour. Immigrants` health and reproductive behaviour however diverge not only from that of the host population, immigrants reproductive health differs also from their population of origin (Milewski 2011). Differences in reproductive health between immigrant women and women in their country of origin however, were neglected in scientific research. Indian women in Austria had moved from a country characterized by high fertility levels to an European country with an exceptionally low fertility rate. Furthermore the social status of women in society differs markedly between Austria and India. In India, women's health, especially reproductive health is associated with their low status in the society (Velkoff and Adlakha 1998, Ravindran and Mishra 2001, Stephenson and Tsui 2002, Aggarwal 2005, Bhargava et al. 2005,Ara et al. 2008,Hazarika 2009, Jayaraman et al. 2009, Mohanty and Pathak 2009, Saikia and Singh 2009, Elul 2011). Just like many centuries back, Indian women mostly still do not have enough rights for themselves, as they still have to live under the control of male family members. Further, most of the health concerns are linked to or aggravated by high levels of fertility. Beside high infant mortality rates (Griffiths et al. 2001, Bhargava et al. 2005), a strong son preference (Pallikadavath and Stones 2006, Puri et al. 2011)motivates the Indian women to bear many children in an attempt to have at least one son survive to adulthood (Ranjana and Awdhesh 2012). Numerous pregnancies, child bearing at a young age and closely spaced gestations and births, however, may affect maternal as well as child health in an adverse manner (Bhalotra and van Soest 2008, Whithworth and Stephenson2002, Rajaram et al. 2009, Santhya et al. 2010). The overall maternal mortality for India is 572.3 per 100 000 births, ranging from 14.9% in Bihar to 1.3% in Kerala (Velkoff and Adlakha 1998). In general women's reproductive health is rather poor in contemporary India. But does the reproductive health situation of Indian women change after migration to Europe? In the present study selected reproductive health parameters of Indian women, who migrated to Austria were focused on for the first time. Additionally selected reproductive health parameters of Indian immigrant women living in Vienna were compared with that of women living in India.

METHODS

Participants and Data Collection

Data collection for this cross-sectional study took place in northern India (Punjab) and in Vienna, Austria between January and April 2010. A series of 107 Indian women aged between 18 and 58 years (x = 36.1 ± 11.3) was enrolled in the present study. Participants were recruited via snowball system and therefore not randomly selected. However, it was no problem to recruit participants because the majority of women were interested in the study and only a few women refused to participate. Two subsamples were defined. Subsample 1 consisted of 57women aged between 18 and 58 years (x = 34.3;

±11.4) living in two villages (Sarai Khas and Kotla) and two cities (Kartarpur and Kapurthala) in the district of Jalandhar in Punjab. Subsample two consisted of 50 women ageing between 19 and 58 years (x = 38.2 ± 10.8) who lived as migrants in Vienna. All of these women originated from Punjab. 7 women (14%) were born in Vienna, 43 women (86%) were born in Punjab. 39 (90.7%) of them migrated to Vienna for family reasons, only 4 (9.3%) reported political or financial reasons. More than 50% rated their German knowledge as very good or good. Data collection in Vienna took mainly place in two temples in Vienna: the Hindu Temple in the 15th district and the Sikh Temple in the 12th district of Vienna. The Indian minority in Vienna is a quite small one. According to Statistik Austria (2009) about 3600 immigrants from India live in Vienna. The Punjabi subgroup is even smaller. For comparison, about 400 000 people with a background of migration, mainly originating from Former Yugoslavia and Turkey, live in Vienna in 2009. As in India, participants were recruited via snowball system. In India as well as in Vienna the aims of the study were explained in detail. Beside the objectives of the study, the right to withdraw at any time was explained. Strict confidential was ensured. The study was conducted in compliance with the "Ethical principles for medical research involving human subjects" of Helsinki Declaration.

Quantitative Data Sampling

Data collection took place by means of face to face interviews carried out by a trained interviewer (M. Singh) based on a standardized structured quantitative questionnaire which was designed by M. Singh for the present study. The questionnaire, which contained 58 questions, was arranged in both languages, German/English for the subjects in Austria as well as English for India. Further, both questionnaires were explained in Hindi and Punjabi for the subjects in Austria and in India. Beside data concerning socioeconomic factors, information about health situation, in particular reproductive health situation were gathered.

Each data collection started with an extensive interview regarding socio-economic parameters such as educational level, professional training, marital status, religion, family size, number of children and place of residence.

The second part of the questionnaire focused on medical history, reproductive history and reproductive health parameters.

Socioeconomic Position (SEP)

Since the most frequently used Hollingshead four factor index of socioeconomic position (SEP) (Hollingshead 1975) was not useful for the present sample, socioeconomic position (low/medium/high) was defined by 7 indicators (educational level, household size, number of rooms per household, educational level of the husband, income level, occupation, occupation of the husband). This system included the three most commonly employed indicators of SEP in contemporary industrialized societies, income, education and occupation (Grundy and Holt 2001; Cirino et al. 2002). A selection of sociodemographic parameters is presented in Table 1. As to be seen in Table 1, not only less educated women of exceptionally low socioeconomic position were enrolled in the present study.

Table 1. Sociodemographic description of the sample (χ^2-test)

		India		Austria		
		n	%	n	%	Sig.
Marital status	Unmarried	17	29.8%	9	18.0%	<0.04
	Married-arranged	36	63.2%	34	68.0%	
	Married not arranged	1	1.8%	5	10.0%	
	widowed	2	3.5%	0	0.0%	
	divorced	1	1.8%	2	4.0%	
Religion	Hinduism	14	24.6%	18	36.0%	<0.02
	Sikhism	39	68.4%	24	48.0%	
	Islam	2	3.5%	8	16.0%	
	Christian	2	3.5%	0	0.0%	
SEP	Low	16	39.0%	6	12.0%	<0.001
	Medium	11	26.8%	38.	76.0%	
	high	14	15.4%	6	12.0%	
Educational level	No school illiterate	14	24.6%	2	4.2%	<0.03
	No school literate	0	0.0%	1	2.1%	
	4 years primary school	7	12.3%	6	12.5%	
	8 years school	12	21.2%	13	27.1%	
	Higher education (college, university)	24	42.1%	26	54.2%	
Employment	yes	16	28.1%	23	46.0%	<0.04
	no	41	71.9%	27	54.0%	

Reproductive Health Parameters

The following reproductive health parameters were collected:

1) Regular visits of a gynecologist (minimum one visit per year), regular visits of a family doctor (minimum two visits per year). Of course the availability of health services are not comparable between a city like Vienna and rural Punjab area. However the access to health services in Vienna, although free of charge, is rather difficult for Indian immigrant women, because no German-Hindi translators are available.
2) Age at first marriage
3) number of children, number of living children
4) age at first birth
5) duration of lactation (in months)
6) history of miscarriage and history of induced abortion
7) Contraceptive use. Contraceptive use means current or previous use of contraceptives such as hormonal contraception and IUR.

Statistical Analyses

Statistical calculations were performed by using SPSS for Windows Program Version 18.0 (Microsoft corp.). After calculation of descriptive statistics (means, SDs), the χ^2-test and the Mann-Whitney-test were applied to test frequency differences with respect to their statistical significance. Multiple and binary logistic regression analyses were calculated to test the impact of migration, age, SEP, religion and educational level on reproductive health parameters. A probability P value of less than 0.05 was considered significant.

RESULTS

Sociodemographic Characteristics

Table 1 presents sociodemographic characteristics of the present sample. Indian women in Punjab and Austria differed significantly in marital status, religion, educational level, socioeconomic position and employment rate. In general educational level and socioeconomic position were higher among women in Austria. Significant differences were also found for marital status. About 68% of subjects in Austria and 63% in India were married. The majority of marriages however were arranged. 10% of women in Austria and only 1.8% in India admitted that they had love marriages and were able to choose their husbands by themselves. Women living in Punjab married significantly earlier than their Austrian counterparts ($p < 0.001$). The average age at first marriage was 20.9 years (±3.7) among women living in India while Indian women in Austria first married at an age of 23.4 years (±2.9). The average years for being married in India were 16.2 years (±10.7) and in Austria 17.5 years (±9.61). The longest period of being married was admitted by a subject in India with 41 years. Punjabi women in Austria were significantly better educated than their Indian counterparts. 90.7% of Punjabi women, who migrated to Austria, did so because of their families. 9.3% of the women living in Austria admitted that they migrated by themselves due to financial reasons.

Reproductive History

Punjabi women in Austria and in India did not differ significantly in the experience of miscarriage, contraceptive use, induced abortions, decision to breastfeed, the regular consultations of a gynaecologist or of a family doctor (see Table 2). Concerning contraceptive use it turned out that hormonal contraceptives prevail (more than 90%), in Austria as well as in Punjab. No women reported the use of emergency contraception (day after pill). Although the availability of health services are not comparable between a city like Vienna and rural Punjab area both subsample did not differ in the frequency of consultation a gynecologist or a family doctor. Most of the subjects reported that they went to a family doctor when they fell ill. But many of them avoided visiting a gynaecologist, except when there was a strong need for it. The allocation was very similar within the two countries. More than 16% of women on Punjab and more than 19% of the Viennese women reported an induced abortion, however we have to state that induced abortion represent a very problematic topic and therefore these data are to interpret with caution. More than 26% of the women were still childless at the time of investigation. This was true of both subsamples. The average amount of living children was in India 1.7 (±1.2), in Austria 1.8 (±1.1). The maximum number of children was 7 in a family in India. Punjabi women in India were significantly younger at the time of first birth (<0.001) (see Table 3). Although nearly 90% of women breastfed their offspring in both subsamples, the duration of lactation differed significantly between Punjabi women in India and Punjabi women in Austria (see Table 3). The maximum breastfeeding years were in India about 7 years (84 months).

Table 2. Reproductive health parameters I - A comparison between Punjabi women in India and inAustria (χ^2-test)

	India		Austria		Significance
Miscarriage	14	25.0%	11	22.4%	n.s.
Previous or current use of contraceptives	18	34.6%	10	22.7%	n.s.
Induced abortion	9	16.1%	9	19.1%	n.s.
Breastfeeding	40	97.6%	33	89.20%	n.s.
Regular visit of a gynaecologist	27	48.2%	21	42.0%	n.s.
Regular visit of a family doctor	42	75.0%	38	76.0%	n.s.

Table 3. Reproductive health parameters II - A comparison between Punjabi women in India and in Austria (Mann-Whitney U-test)

	India		Austria		Significance
	x (SD)	range	x (SD)	range	
Number of children	2.2 (1.7)	0-7	1.8 (1.4)	0-5	n.s
Number of living children	1.7 (1.2)	1-6	1.8 (1.1)	1-5	n.s.
Number of desired children	1.3 (1.2)	0-3	1.8 (1.1)	0-4	n.s.
Age at first marriage	20.9 (3.7)	15-29	23.4 (2.9)	17-30	<0.001
Age at first birth	21.2 (3.2)	17-30	24.5 (2.8)	20-30	<0.001
Duration of lactation (months)	19.5 (17.1)	0-84	8.4 (8.9)	0-48	<0.001

Impact of Sociodemographic Factors on Reproductive Health

Table 4 highlights the results of the multiple regression analyses. As to be seen, age, country of living, and educational level had a significant impact the number of children. The number of still living children was significantly related to age and educational level. Age at first birth increased significantly with educational level and was related to the country of living. As to be seen in Table 5 the use of contraceptives and the duration of lactation were significantly influenced by educational level. The use of contraceptives was also influenced by the country of living. In general, with increasing educational level the number of offspring and still living offspring but also the duration of lactation decreased. On the other hand an increasing educational level increased the age at first birth significantly. Furthermore educational level had a positive impact on the use of contraceptives (see Tables 4 and 5).

Table 4. Impact of sociodemographic factors on reproductive health parameters. Multiple regression analyses

	Multiple R	Regression coefficient B	Significance	95% confidence interval
Dependent variable: Number of children				
constant	0.78	0.29	n.s.	-0.88 -1.46
Age		0.09	<0.001	0.07-0.11
Country (India/Austria)		-0.81	<0.001	-1.23- -0.39
Educational level		-0.45	<0.001	-0.77 –0.12
Socioconomic position (SEP)		0.05	n.s.	-0.13 -0.22
Religion		0.77	n.s.	-0.21 – 0.36
Dependent variable: Number of children living today				
constant	0.61	0.67	n.s.	-0.52- 1.86
Age		0.05	<0.001	0.03 – 0.08
Country (India/Austria)		-0.25	n.s.	-0.71 – 0.21
Educational level		-0.08	<0.05	-0.26 – 0.11
Socioconomic position (SEP)		-0.13	n.s.	-0.51 -0.24
Religion		0.25	n.s.	-0.03 -0.53
Dependent variable: age at first marriage				
constant	0.51	16.76	<0.0001	12.66 – 20.86
Age		-0.02	n.s.	-0.08 – 0.04
Counrty (India/Austria)		1.49	<0.05	-0.09 – 3.07
Educational level		0.79	<0.01	0.19 -1.38
Socioeconomic position		0.32	n.s.	-0.91 – 1.55
Religion		0.34	n.s.	-0.68 – 1.36
Dependent variable: Age at first birth				
constant	0.52	17.29	<0.000	13.24 -21.36
Age		0.02	n.s.	-0.06 -3.21
Country (India/Austria)		1.61	<0.04	0.01 -3.21
Educational level		0.74	<0.02	0.11 -1.37
Socioconomic position (SEP)		0.26	n.s.	-1.02 -1.55
Religion		-0.27	n.s.	-1.24 – 0.71
Dependent variable: duration of lactation				
constant	0.58	3.17	<0.001	1.73 -4.61
Age		0.02	n.s.	-0.01 -0.05
Country (India/Austria)		-1.03	<0.001	-0.01 -0.05
Educational level		-0.18	n.s.	-0.41 – -0.05
Socioconomic position (SEP)		-0.19	n.s.	-0.65 -0.28
Religion		0.01	n.s.	-0.34 – 0.36

Table 5. Impact of sociodemographic factors on reproductive health parameters. Binary regression analyses

	Regression coefficient B	Significance	95% confidence interval
Dependent variable:History of miscarriage (no, yes)			
constant	-1.93	n.s.	
Age	0.04	n.s.	0.98 -1.09
Country (India/Austria)	-0.18	n.s.	0.28 -2.51
Educational level	-0.15	n.s.	0.56 – 1.32
Socioconomic position (SEP)	0.07	n.s.	0.46 -2.51
Religion	-0.02	n.s.	0.47 -3.03
Dependent variable:Induced abortion (no, yes)			
constant	-5.16	<0.01	
Age	0.04	n.s.	0.98 -1.11
Country (India/Austria)	-0.07	n.s.	0.27 -3.27
Educational level	0.34	n.s.	0.79 -2.49
Socioconomic position (SEP)	0.29	n.s.	0.48 -3.64
Religion	0.14	n.s.	0.51 -2.55
Dependent variable: Use of contraceptives (no, yes)			
constant	-0.55	n.s.	
Age	-0.03	n.s.	0.92 – 1.02
Country (India/Austria)	-1.44	<0.01	0.07 -0.76
Educational level	1.09	<0.02	1.18 -7.43
Socioconomic position (SEP)	0.23	n.s.	0.75 -2.11
Religion	-0.01	n.s	0.45 -2.16
Dependent variable: Regular consultation of a gynaecologist (no, yes)			
constant	-1.95	n.s.	
Age	0.04	n.s.	0.99 -1-09
Country (India/Austria)	-0.75	n.s	0.17 -1.29
Educational level	0.37	n.s.	0.95 – 2.21
Socioconomic position (SEP)	0.32	n.s.	0.64 -2.93
Religion	-0.36	n.s.	0.36 -1.36
Dependent variable: Lactation (no, yes)			
constant	-2.46	n.s.	
Age	0.28	n.s.	0.98 – 1.79
Country (India/Austria)	-3.98	n.s.	0.01 -1.76
Educational level	-1.06	<0.05	0.14 – 0.84
Socioconomic position (SEP)	1.25	n.s.	0.09 – 1.36
Religion	-0.58	n.s.	0.07 – 4.84

DISCUSSION

Recent immigration to Western European countries is largely from countries with high fertility rates to countries with exceptionally low fertility rates. Female immigrants differ consequently in fertility patterns and reproductive health behaviour from their country of destination. In comparison to the host population immigrants are mainly characterized by early entrance into marriage, early childbearing and a higher number of offspring and a reduced use of contraceptives. Immigrants originating from rural areas however, are also confronted with culture typical norms of their mostly urban host population and undergo a process of assimilation and adaptation. Fargues (2006) claims that most of recent international migration has been from high to low fertility rate countries and that migrants typically adopt reproductive behavior patterns of the host population. On the other hand Fernandez (2007) pointed out, that migrants tend to carry on their indigenous culture with them, even along generations. This is especially true of large immigrant communities which create their own cultural setting with a low degree of assimilation and integration such as the Turkish immigrant community in Western Europe (Milewski 2011). Previous research predominantly focused on these large immigrant communities, but what about small immigrant communities?

Up to now no study had focused on the reproductive health situation of Indian immigrant women in Austria in comparison to women in India. This is mainly due to the fact that the Indian minority group is a very small one in Austria. Consequently the present study is the very first one which focused on the reproductive health situation of Indian women originating from Punjab, who migrated to Vienna, Austria. Before we start to discuss the results in detail, we have to state that the present study has certain limitations. One shortcoming is the small sample size (n = 107), which results mainly from the few potential participants in Vienna. According to Statistik Austria (2009), about 3600 immigrants from India live in Vienna. The Punjabi subgroup is even smaller, consisting of few hundreds of persons however, no reliable census data exist. The other problem is the cross-sectional design, which allows limited interpretations only.

The results of the present study show that the Indian women enrolled in the present study in Vienna, Austria, differ markedly in some reproductive health parameters from their counterparts in Punjab. Punjabi women in Austria married significantly later and were significantly older when they gave first birth than their counterparts in India. Mean age at first birth was 21.7 years in India, while Punjabi women in Vienna were 24.5years old on the average when they gave first birth. This increase in the age at first marriage and the delay of childbearing, may be as a result of later marriage, among immigrant women in comparison to their population of origin was also described for Turkish and Moroccan women in Western Europe (Schoenmaeckers et al.

1999, Milewski 2011). The delay in childbearing in immigrant women may be also a result of longer and improved education, changed family situation and the wish to adapt to the fertility patterns of the host population (Milewski 2011). Nevertheless the average age at first birth of the Austrian host population is significantly higher than among Punjabi immigrant women. The mean age at first birth among Austrian women is 29.7 years (Statistik Austria 2009). Consequently Punjabi women in Austria seem to represent an intermediate stage between Indian and Austrian reproductive behaviour. However we have to be aware that Indian women of the present sample had not finished their reproductive carrier at the time of data collection. However, Indian women delayed clearly age at first marriage and age at first birth in comparison to their counterparts in India. This trend to postpone family formation corresponds to a move toward a more western behaviour which was already reported for Turkish and Moroccan women (Schoenmaeckers et al. 1999). Regarding mean number of offspring no significant differences between Indian immigrant women in Austria and women in Punjab were found. This finding is in contrast to the results of previous studies which reported a slight decline in the number of offspring among other immigrant groups in Western Europe (Schoenmaeckers et al. 1999, Milewski 2011). Changes in reproductive behaviour among immigrant women, in particular the delay of first marriage and first birth and a reduced number of offspring was mainly explained within the framework of assimilation theory (Stephan and Bean 1992) or it was interpreted as an example of adaptive behaviour (Landale and Hauan 1996). According to this viewpoint immigrant women try to adapt to reproductive behaviour of the host population, as already described by Schoenmaeckers et al. (1999), Fargues (2006) and Milewski (2011). Nevertheless the average number of offspring among Austrian women (x = 1.39 children) is quite lower than the average number of offspring of Punjabi women in Vienna of the present sample (x = 1.81 children). In this context we have to state again that the majority of Indian women in the present sample had not finished their reproductive carrier at the time of the study, while 1.39 children on average is typical of Austrian women who had already finished their reproduction. Another effect of assimilation towards reproductive behaviour typical of the host country seems to be the reduced length of breastfeeding found among Punjabi women in Vienna in comparison to their counterparts in India. 81% of the Indian women in Austria and 87.0% of the women India reported to breastfeed their offspring. The women in India however, reported a significantly longer duration of lactation. While Indian women in Punjab breastfed their offspring for 19.5 months on the average, among Punjabi women in Austria the mean duration of lactation decreased to 8.4 months. Breastfeeding patterns and the duration of lactation underlie cultural norms and differ by ethnicity (De Amici et al. 2001). Migration seems to influence breastfeeding patterns too. A reduced period of breastfeeding was reported for various groups of immigrants such as Ethiopian immigrant women in Israel (Rubin 2010). Immigration results often drastically reduced breastfeeding periods (Kannan et al. 1999, Twamley et al. 2011).

We can summarize that migration from rural Punjab to the Austrian capital Vienna is associated with a postponement of family formation, but also in decreased duration of lactation. These effects may be interpreted as a part of acculturation to urban Austrian fertility habits and corroborate the hypothesis of Fargues (2006). On the other hand Indian immigrant women gave birth to markedly more offspring than the Austrian host population, an observation which corroborates the results of Fernandez (2007). Consequently Punjabi women in Vienna undergo a reproductive transition from the high fertility of their population of origin towards the low fertility of their country of destination, but have not yet reached the low fertility level of Austrian women. Indian women in Austria live in a very small immigrant community. They are exposed to values and norms typical of their own culture but they are also confronted with completely different norms and values of their host population. They are significantly better educated than their counterparts in India and they are significantly more often employed and earn their own money. Better and longer education is the first step to fertility transition and assimilation and results in a delay of first marriage and first childbearing and consequently in a lower number of offspring (Islam and Nesa 2009). The present study revealed that educational level was the most important factor influencing reproductive health parameters among Indian women in Austria and in Punjab. With increasing educational level number of offspring decreased and the age at first birth and first marriage increased. Furthermore educational influenced the use of contraceptives, health related behaviour and the duration of lactation. These findings are in accordance with those of other studies, which reveal that maternal educational level has a positive impact on birth weight, child mortality, use of contraceptives and general health (Velkoff and Adlakha 1998, Olson and Madsen 1999, Casterline et al. 2001, Aggarwal 2005, Mohanty and Pathak 2009, Saikia and Singh 2009, Ergin et al. 2010).

Punjabi women in Vienna however, did not differ in other reproductive health related behaviors, such as contraceptive use, consultation of a gynecologist, from Indian women in Punjab. This finding may be interpreted as a result of social disadvantage of immigrant women in Austria. Although their educational level is significantly higher than among women in India and many of Punjabi women in Vienna work and earn their own money, they are still a disadvantaged group In Austria because they are immigrants. Consequently some Indian migrant women refused to visit a gynecologist although medical treatment is free of charge in Austria. This problem of gender equity and socioeconomic inequality among immigrant women and its effects on women's health is not only a problem of Austria (Moss 2004, Llacer et al. 2007). The adverse effects of gender and socioeconomic disadvantage on reproductive health were pointed out by some studies (Brotto et al. 2008, Puri et al. 2011). Therefore countries of immigration, such as Austria should include these aspects of gender and socioeconomic disadvantage of immigrant women in the public health strategies.

Acknowledgments

The authors are gratefully indebted to their volunteers for their patience and cooperation.

References

Abate, Nicola, and Manisha Chandalia. 2007. "Ethnicity, type 2 diabetes and migrant Asian Indians." *Indian Journal of Medical Research* 125:251-258.

Adanu, RichardM. K., and Timothy R. B. Johnson. 2009. "Migration and women's health." *International Journal of Gynecology and Obstetrics* 106:179-181.

Aggarwal, RashmiK.2005. "Conventional and oral contraceptive users in rural Punjab." *Journal of Family Welfare* 51:1-9.

Amici de, Donatella, Gasparoni, Antonella, Guala, Andrea, and Catherine Klersy. 2001. "Does ethnicity predict lactation? A study of four ethnic communities." *European Journal of Epidemiology* 17:357-362.

Ara, Gulshan, Siddique, YasirH., Tanveer, Beg, and Mohammad Afzal. 2008. "Abortion index and mortality of offspring among women of different age, caste and population groups of north Indian Muslims." *Journal Biosocial Science* 40: 431-443.

Bhalotra, Sonia, and Anthur van Soest. 2008." Birth spacing, fertility and neonatal mortality in India, Dynamics, frailty and fecundity." *Journal of Econometrics* 143:274-290.

Bhargava, Alok, Chowdhury, Sadia, and Kumal K. Singh.2005. "Healthcare infrastructure, contraceptive use and infant mortality in Uttar Pradesh, India." *Economics and Human Biology*3:388-404.

Bollini, Paola, Pampallona, Sandro, Wanner, Philippe, and Bruce Kupelnick. 2009. "Pregnancy outcome of migrant women and integration policy, A systematic review of the international literature." *Social Sciences and Medicine* 68:452-461.

Bongard, Stephan, Pogge, Sandra F., Arslaner, Halime, Rohrmann, Sonja, and Vilker Hodapp. 2002. "Acculturation and cardiovascular reactivity of second generation Turkish migrants in Germany." *Journal of Psychosomatic Research*53:795-803.

Brotto, Lori, Chou, Anni, Singh, Tamo, and Jane Woo. 2008. "Reproductive Health practices among Indian, Indo-Canadian, Canadian East Asian and Euro-Canadian women, the role of acculturation." *Journal of Obstetrics and Gynecology Canada* 30:229-238.

Bundey, Sarah, Alam, Hasina, and Amritpal Kaur.1991. " Why do UK-born Pakistani babies have high perinatal and neonatal mortality rates?" *Paediatric and Perinatal Epidemiology* 5:101-114.

Casterline, John B., Sathar A., Zeba, and MinhajHaque. 2001. "Obstracles to contraceptive use in Pakistan, A study in Punjab." *Studies in Family Planning* 32:95-110.

Cirino, Paul T., Chin, Christopher E., Sevcik, Rose A., Wolf, Maryanne, Lovett, Maureen, and Robin D. Morris. 2002." Measuring socioeconomic status: reliability and preliminary validity for different approaches." *Assessment* 9:145-155.

Daryani, Achraf, Berglund, Lars, Andersson, Asa, Kocturk, Tahire, Becker, Wulf, and Bengt Vessby. 2005. "Risk factors for coronary heart disease among immigrant women from Iran and Turkey compared to women of Swedish ethnicity." *Ethnicity Disease*15:213-220.

Deville, Walter, Greacen, Tim, Bogic, Marija, Dauvrin, Marie, Dias, Sonia, Gaddini, Andrea, Natasja, Koitzsch Jensen, Karamanidou, Christiana, Kluge, Ulrike, Merzaniemi, Ritva, Puigpinos Riera, Rosa, Sarvary, Attila, Soares, Joaquim J. F., Stankunas, Mindaugas, Straßmayr, Christa, Welbel, Marta, and StefanPriebe.2011. "Health care for immigrants in Europe: Is there still consensus among country experts about principles ogf good practice? A Delphi study." *Public Health* 11:699-679.

Dotevall, Annika, Rosengren, Annika, Lappas, Georg, and Lars Wilhelmsen. 2000. "Does immigration contribute to decreasing CHD incidence? Coronary risk factors among immigrants in Göteborg, Sweden." *Journal of International Medicine* 247:331-339.

Elul, Batya. 2011. "Determinants of induced abortion: an analysis of individual, household and contextual factors in Rajasthan." *Journal of Biosocial Science*11:43:1-17.

Ergin, Isil, Hassoy, Hur, Tanik, Feride A., and Gokce Aslan.2010. "Maternal age, educational level and migration, socioeconomic determinants for smoking during pregnancy in a field study from Turkey." *Public Health* 10:325-334.

Fargues, Philippe. 2006. *The demographic benefit of migration: hypothesis and application to Middle Eastern and North African contexts*. World Bank Policy Research Working Paper 4050, World Bank.

Fernadnez, Raquel. 2007. "Women, work and culture." *Journal of the European EconomicAssociation* 5:305-332.

Hawkins, Summer S., Lamb, Kate, Cole, Tim J., and Catherine Law. 2008. "Influence on moving to the UK on maternal health behaviors: prospective cohort study." *British Medical Journal* 336:1052-1055.

Hazarika, Indrajit. 2009. "Women's reproductive health in slum populations in India: Evidence from NFHS-3." *Journal of Urban Health* 87:264-277.

Hollingshead, August B. 1975. *Four factor index of social status*. New Heaven: Yale University, CT.

Griffiths, Paula, Hinde, Andrew, and Zoe Matthews. 2001. " Infant and child mortality in three culturally contrasting states of India." *Journal of Biosocial Science*33:603-622.

Grundy, Emily, and Grady Holt. 2001. "The socioeconomic status of older adults: How should we measure it in studies of health inequalities?" *Journal of Epidemiology and Community Health* 55:895-904.

Iglesias, Edgar, Robertson, Eva, Johanson, Seven-Erik, Engfeldt, Peter, and Jan Sundquist. 2003. "Women, international migration and self-reported health. A population-based study of women of reproductive age." *Social Science and Medicine*56:111-124.

Islam, Sabina, and Mossamet K. Nesa.2009. "Fertility transition in Bangladesh: the Role of education." *Proceedings Pakistan Academy of Sciences* 46:195-201.

Jayaraman, Anuja, Mishra, Vinod, and Fred Arnold. 2009. "The relationship of family size and composition to fertility desires, contraceptive adoption, and method choice in south Asia." *International Perspectives on Sexual and Reproductive Health* 35:29-38.

Jonkers, Marina, Richters, Annemiek, Zwart, Joost, Öry, Ferko, and Jos van Roosmalen. 2011. "Severe maternal morbidity among immigrant women in the netherlands, patients perspectives." *Reproductive Health Matters*; 19:144-153.

Kannan, Srimathi, Carruth, Betty R., and Jean Skinner. 1999. "Infant feeding practices of Anglo-American and Asian Indian American mothers." *Journal of American College of Nutrition* 18:279-286.

Landale, Nancy S., and Susan M. Hauan.1996. "Migration and premarital childbearing among Puerto Rican women." *Demography* 33:429-442.

Landman, Joseph, and John Cruickshank. 2001. "A review of ethnicity, health and nutrition related diseases in relation to migration in the United Kingdom." *Public Health Nutrition* 4:647-657.

Llacer, Alicia, Zunzunegui, Miaria V., del Amo, Julia, Mazarrasa, Lucia, and FranciscoBolumar. 2007. "The contribution of a gender perspective to the understanding of the migrant'shealth." *Journal of Epidemiology and Community Health* 61:4-10.

Madan, Ashima, Palaniappan, Latha, Urizar, Guido, Wang, Yun, Fortman, Stephen P., and Jeffrey B. Gould.2006. "Sociocultural factors that affect pregnancy outcomes in two dissimilar immigrant groups in the United States." *Journal of Pediatrics* 148:341-346.

Meadows, Lynn M., Thurston, Wilfreda E., and ChristianaMelton.2001. " Immigrant women's." *Social Science and Medicine* 52:1451-1458.

Milewski, Nadja. 2011. "Transition to a first birth among Turkish second-generation migrants in Western Europe." *Advances in Life Course Research* 16:178-189.

Minnis, Alexandra M. 010. "US migration and reproductive health among Mexican women, assessing the evidence for health selectivity." *Facts Reports* 2:1-7.

Misra, Anoop, and Om P. Ganda.2007. "Migration and its impact on adiposity and type 2 diabetes." *Nutrition*23:696-708.

Mohanty, Sanjay, and Praveen Pathak. 2009. "Rich-poor gap in utilization of reproductive and child health services in India, 1992-2005." *Journal of Biosocial Science* 41:381-398.

Moss, Nancy. 2002. "Gender equity and socioeconomic inequality, a framework for the patterning of women's health." *Social Science and Medicine* 54:649-661.

Muynck, Aime. 1997. "Migration and Health in Belgium." *InCountry Reports on Migration and Health in Europe*, edited by A. Huismann, Caren Weilandt, and A. Geiger. Bonn:Wissenschaftliches Institut der Ärzte.

Olson, Ole, and Mette Madsen. 1999. "Effects of maternal education on infant mortality and stillbirths in Denmark." *Scandinavian Journal Public Health* 27:128-136.

Pallikadavath, Saseendran, and William R. Stones. 2006. "Maternal and social factors associated with abortion in India, a population-based study." *International Family Planning Perspectives* 32:120-125.

Puri, Sunita, Adams, Vincanne, Ivey, Susan, and Robert D. Nachtigal. 2011. "There is such a thing as too many daughters, but not too many sons: A qualitative study of son preference and fetal sex selection among Indian immigrants in the United States." *Social Science and Medicine* 72:1169-1176.

Rajaram, Shyama, Zottarelli, Liza K., and Thankam S. Sunil. 2009. "An assessment of fetal loss among currently married women in India." *Journal of Biosocial Science* 41:309-327.

Ranjana, Kesawani, and Yadav Awdhesh. 2012. "Role of Gender in contraceptive use among currently married women in Uttar Pradesh and Bihar." *Research on Humanities and Social Sciences* 2:34-44.

Ravindran, Sundari, and UsMishra. 2001. "Unmet need for reproductive health in India." *Reproductive Health Matters* 9:105-113.

Rubin, Lisa, Nir-Inbar, Sharon, and Shmuel Rishpon. 2010. "Breastfeeding patterns among Ethiopianimmigrant mothers, Israel, 2005-2006." *Israel Medical Association Journal*12:657-661.

Saikia, Nandita, and Abhishek Singh. 2009. "Does type of household affect maternal health? Evidence from India." *Journal of Biosocial Science* 41:329-353.

Santhya, K. G., Ram, Usha, Acharya, Rajib, Jejeebhoy, Shireen J, Ram, Faujdar, and Abhishek Singh. 2010. "Associations between early marriage and young women's marital and reproductive health outcomes: evidence from India." *International Perspectives on Sexual and Reproductive Health* 36:132-139.

Sargent, Carolyn, and Stephanie Larchanche. 2011. "Transnational migration and global health: The production and management of risk, illness and access to care." *Annual Review in Anthropology* 40:345-361.

Singh, Maryam, and Sylvia Kirchengast. 2011. "Obesity prevalence and nutritional habits among Indian women: a comparison Between Punjabi women living in India and Punjabi migrants in Vienna, Austria." *Anthropologischer Anzeiger* 68:239-51.

Schoenmaeckers, Ronald C., Lodewijckx, Edith, and SylvieGadeyne. 1999. "Marriages and fertility among Turkish and Moroccan women in Belgium: results from census data." *International Migration Review* 33:901-928.

Statistik Austria. 2009. *Statistisches Jahrbuch.* Österreich Verlag Österreich, Austria.

Stephen, Elizabeth, and Frank Bean. 1992. "Assimilation, disruption and the fertility of Mexican-origin women in the United States." *International Migration Review*26:67-88.

Stephenson, Rob, and AmyTsui O. 2002. "Contextual influences on reproductive health service use in Uttar Pradesh." *Studies on Family Planning* 33:309-320.

Twamley, Katherine, Puthussery, Shuby, Harding, Seeromanie, Baron, Maurina, and Alison Macfarlane. 2011. "UK-born ethnic minority women and their experiences of feeding their newborn infant." *Midwifery* 27:595-602.

Uraquia, Marcelo L., Glazier, Richard H., Blondel, Beatrice, Zeitlin, Jennifer, Gissler, Mika, Macfarlane, Alison, Ng, Edward, Heamean, Maureen, Stray-Pedersen, Babill, and AnitaJ. Gagnon. 2010. "International migration and adverse birth outcomes, role of ethnicity, region of origin and destination." *Journal of Epidemiology and Community Health* 64:243-251.

Ujcic-Voortman, Joanne, K., Schram, Miranda T., Jacobs-van der Bruggen, MiniqueA., Verhoeff, Amound P., and Caroline A. Baan.2009. "Diabetes prevalence and risk factors among ethnic minorities." *European Journal of Public Health* 19:511-515.

Uitewaal, Paul J. M., Manna, Dyno R., Bruijnzeels, Mark, Hoes, Arno W., and SabuThomas. 2004. "Prevalence of type 2 diabetes mellitus, other cardiovascular risk factors, and cardiovascular disease in Turkish and Moroccan immigrants in North West Europe: a systematic review." *Preventive Medicine* 39:1068-1076.

Velkoff, Victoria, and Arjun Adlakha. 1998. *Women's health in India*. US Department of Commerce, Bureau of Census.

Whitworth, Alison, and Rob Stephenson.2002. "Birth spacing, sibling rivalry and child mortality in India." *Social Science and Medicine* 55:2107- 2119.

EDITORS' CONTACT INFORMATION

Dr. Slawomir Koziel
Director, Polish Academy of Sciences
Anthropology Unit in Wroclaw,
ul. Podwale 75, 50-449, Wroclaw, Poland.
slawomir.koziel@antropologia.pan.pl

Dr. Raja Chakraborty
Head of Department,
Department of Anthropology,
Dinabandhu Mahavidyalaya,
Bongaon, West Bengal, India.
rajanth2003@yahoo.co.uk

Dr. Kaushik Bose
Professor.
Department of Anthropology,
Vidyasagar University,
Midnapore, West Bengal, India
banda@vsnl.net and kaushikbose@cantab.net

INDEX

C

D

E

F

G

H

I

N

O

P

T

U

V

W

Y